Springer Tracts in Advanced Robotics

Volume 153

Frank Chongwoo Park, Mechanical Engineering Department, Seoul National University, Seoul, Korea (Republic of)

S. E. Salcudean, The University of British Columbia, Vancouver, BC, Canada

Roland Siegwart, Institute of Robotics and Autonomous Systems Lab, ETH Zürich, Zürich, Switzerland

Gaurav S. Sukhatme, Department of Computer Science, University of Southern California, Los Angeles, CA, USA

The Springer Tracts in Advanced Robotics (STAR) publish new developments and advances in the fields of robotics research, rapidly and informally but with a high quality. The intent is to cover all the technical contents, applications, and multi-disciplinary aspects of robotics, embedded in the fields of Mechanical Engineering, Computer Science, Electrical Engineering, Mechatronics, Control, and Life Sciences, as well as the methodologies behind them. Within the scope of the series are monographs, lecture notes, selected contributions from specialized conferences and workshops, as well as selected Ph.D. theses.

Special offer: For all clients with a print standing order we offer free access to the electronic volumes of the Series published in the current year.

Indexed by SCOPUS, DBLP, EI Compendex, zbMATH, SCImago.

All books published in the series are submitted for consideration in Web of Science.

Antonio Loquercio

Agile Autonomy: Learning High-Speed Vision-Based Flight

 Springer

Antonio Loquercio
University of Zurich
Zürich, Switzerland

ISSN 1610-7438 ISSN 1610-742X (electronic)
Springer Tracts in Advanced Robotics
ISBN 978-3-031-27287-5 ISBN 978-3-031-27288-2 (eBook)
https://doi.org/10.1007/978-3-031-27288-2

This Springer imprint is published by the registered company Springer Nature Switzerland AG
The registered company address is: Gewerbestrasse 11, 6330 Cham, Switzerland

To my advisor,
who showed me what robots can learn.
and
To my family, Noëmi, and friends,
who showed me what robots cannot learn.

Series Editor's Foreword

At the dawn of the century's third decade, robotics is reaching an elevated level of maturity and continues to benefit from the advances and innovations in its enabling technologies. These all are contributing to an unprecedented effort to bringing robots to human environment in hospitals and homes, factories, and schools; in the field for robots fighting fires, making goods and products, picking fruits, and watering the farmland, saving time and lives. Robots today hold the promise for making a considerable impact in a wide range of real-world applications from industrial manufacturing to healthcare, transportation, and exploration of the deep space and sea. Tomorrow, robots will become pervasive and touch upon many aspects of modern life.

The *Springer Tracts in Advanced Robotics (STAR)* is devoted to bringing to the research community the latest advances in the robotics field on the basis of their significance and quality. Through a wide and timely dissemination of critical research developments in robotics, our objective with this series is to promote more exchanges and collaborations among the researchers in the community and contribute to further advancements in this rapidly growing field.

The monograph by Antonio Loquercio is based on the author's doctoral thesis. The contents are focused on aerial robotics and are organized in three chapters. Robust perception and action techniques are proposed for high-speed quadrotor flight in unstructured environments with only onboard sensing and computation. The book also presents novel ways to estimate uncertainty in neural network prediction which is promising for the development of data-driven algorithms.

Rich of examples developed by means of simulation and extensive experimentation on different flying platforms, this volume was the winner of the 2022 Georges Giralt Ph.D. Award for the best doctoral thesis in Europe. A very fine addition to the STAR series!

Naples, Italy
March 2023

Bruno Siciliano
STAR Editor

Acknowledgements

First of all, I would like to thank my supervisor Prof. Davide Scaramuzza, who accepted me as a Ph.D. student and provided me with many exciting opportunities and advice over the years. I would also like to particularly thank my colleague Elia Kaufmann, without whom the majority of the work in this thesis would not exist. Elia always supported me and helped me to transform my crazy ideas into reality.

I wish to express my gratitude to all the current and past members, visitors, and students who I encountered during my journey as a Ph.D. student at the Robotics and Perception Group. I would particularly like to thank Matthias Fässler, Henri Rebecq, Davide Falanga, Titus Cieslewski, Zichao Zhang, Philipp Föhn, Mathias Gehrig, Daniel Gehrig, Manasi Muglikar, Yunlong Song, Giovanni Cioffi, Nico Messikommer, Jeff Delmerico, Suseong Kim, Guillermo Gallego, Dario Brescianini, Dimche Kostadinov, Javier Hidalgo Carrio, Christian Pfeiffer, Julien Kohler, Thomas Längle, Manuel Sutter, Kevin Kleber, Kunal Shrivastava, Stefano Ghidoni, Rubén Gómez Ojeda, Naveen Kuppuswamy, Timo Stoffregen, Francisco Javier Perez Grau, Bianca Sangiovanni, Yuto Suebe, Roberto Tazzari, Sihao Sun, Ana Maqueda, and Tamar Tolcachier. I also had the pleasure to work with great students, namely, Mario Bonsembiante, Lorenzo Ferrini, Simone Arreghini, Alessandro Saviolo, Francesco Milano, Daniel Mouritzen, Bojana Nenezic, Yawei Ye, Mattia Segu, Christoph Meyer, Simon Muntwiler, and Moritz Zimmermann.

Furthermore, I was fortunate enough to work with great international collaborators, namely, Yanchao Yang, Stefano Soatto, Alexey Dosovitskiy, Rene Ranftl, Matthias Muller, Vladlen Koltun, Daniele Palossi, Francesco Conti, Eric Flamand, Luca Benini, Antoni Rosinol, and Luca Carlone. I would like to thank the agencies funding my research, namely, Intel, the National Centre of Competence in Research (NCCR) Robotics, the Swiss National Science Foundation, and the European Research Council.

I would like to thank Prof. Angela Schoellig, Prof. Pieter Abbeel, and Prof. Roland Siegwart for accepting to review my thesis.

Last but not least, I am very grateful to Noëmi, my family, and my friends without whom this work would make no sense.

Zürich, Switzerland   Antonio Loquercio
July 2021

About This Book

"The robot revolution has arrived", says the stunning title of a 2020 article in the National Geographic. Over the course of the last decades, autonomous robots have had a tremendous impact on our global economy. Indeed, machines now perform all sorts of tasks: they take inventory and clean floors in big stores; they shelve goods and fetch them for mailing in warehouses; they patrol borders; and they help children with autism. In the majority of those applications, they co-exist and interact with their surroundings, being that humans or other autonomous systems, to create an ecosystem that maximizes efficiency and productivity.

Yet, a major challenge for autonomous systems to operate in unconstrained settings is to cope with the constant variability and uncertainty of the real world. This challenge currently limits the application of robots to structured environments, where they can be closely monitored by expensive suites of sensors and/or human operators. One way to tackle this challenge is exploiting the synergy between perception and action which characterizes natural and artificial agents alike. For embodied agents, action controls the amount of information coming from sensory data, and perception guides and provides feedback to action. While this seamless integration of sensing and control is a fundamental feature of biological agents [55], I argue that a tight perception-action loop is fundamental in enriching the autonomy of artificial agents.

My thesis investigates this question in the context of high-speed agile quadrotor flight. Given their agility, limited cost, and widespread availability, quadrotors are the perfect platform to demonstrate the advantages of a tightly coupled perception and action loop. To date, only human pilots can fully exploit their capabilities in unconstrained settings. Autonomous operation has been limited to constrained environments and/or low speeds. Having reached human-level performance on several tasks, data-driven algorithms, *e.g.*, neural networks, represent the ideal candidate to enhance drones' autonomy. However, this computational tool has mainly proven its impact on disembodied datasets or in very controlled conditions, *e.g.*, in standardized visual recognition tasks or games. Therefore, exploiting their potential for high-speed aerial robotics in a constantly changing and uncertain world poses both technical and fundamental challenges.

This thesis presents algorithms for tightly coupled robotic perception and action in unstructured environments, where a robot can only rely on onboard sensing and computation to accomplish a mission. In the following, we define a policy as tightly coupled if it directly goes from onboard sensor observations to actions without intermediate steps. Doing so creates a synergy between perception and action, and it enables end-to-end optimization to the downstream task. To design such policy, I explore the possibilities of data-driven algorithms. Among others, this thesis provides contributions to achieving high-speed agile quadrotor flight in the wild, pushing the platform closer to its physical limits than what was possible before. Specifically, I study the problem of navigation in natural and man-made (hence in the wild) previously unknown environments with only onboard sensing and computations (neither a GPS nor external links are available to the robot). To achieve this goal, I introduce the idea of sensing and action abstraction to achieve knowledge transfer between different platforms, e.g., from ground to aerial robots, and domain, e.g., from simulation to the real world. This thesis also presents novel ways to estimate uncertainty in neural network predictions, which open the door to integrating data-driven methods in robotics in a safe and accountable fashion. Finally, this thesis also explores the possibilities that the embodied nature of robots brings to tackle complex perception problems, both for low-level tasks such as depth estimation or high-level tasks (e.g., moving object detection). Overall, the contributions of this work aim to answer the question *Why learning tightly coupled sensorimotor policies in robotics?* In the following is a list of contributions:

- The introduction of data-driven algorithms to the problem of high-speed agile quadrotor flight in unstructured environments with only onboard sensing and computation.
- A novel architecture for sensorimotor control of a drone, which is trained only with data collected from ground platforms, e.g., cars and bicycles. This method allows navigation in both indoor and outdoor environments and is robust to dynamic obstacles.
- A theoretically motivated method based on abstraction to transfer knowledge between platforms and domains, e.g., from simulation to the physical world.
- The application of this method to train deep sensorimotor policies for drone racing, acrobatics, and navigation in the wild, demonstrating that a policy trained entirely in simulation can push robots close to their physical limits.
- A general and theoretically motivated framework to estimate the uncertainty of neural network predictions, which is demonstrated on several perception and control tasks.
- A framework for unsupervised training of neural networks on the task of moving object detection. The framework only uses video data and its statistical properties to predict moving objects.

- A novel neural architecture and an associated training procedure to learn 3D perception from monocular image data. The architecture is trained with only the supervision that a robot would observe while exploring a scene: images and very sparse depth measurements.

Contributions

Journal Publications

- **Antonio Loquercio**, Elia Kaufmann, Rene Ranftl, Matthias Mueller, Vladlen Koltun, and Davide Scaramuzza. "Agile Autonomy: Learning High-Speed Flight in the Wild". *Science Robotics* (2021).
  DOI: 10.1126/scirobotics.abg581
  Links: Code
- **Antonio Loquercio**, Alexey Dosovitskiy, and Davide Scaramuzza. "Learning Depth with Very Sparse Supervision". In: *IEEE Robotics and Automation Letters (RA-L)* (2020).
  DOI: 10.1109/LRA.2020.3009067
  Links: PDF, Code
- **Antonio Loquercio**, Mattia Segu, and Davide Scaramuzza. "A General Framework for Uncertainty Estimation in Deep Learning". In: *IEEE Robotics and Automation Letters (RA-L)* (2020).
  DOI: 10.1109/LRA.2020.2974682
  Links: PDF, Video, Code
- **Antonio Loquercio**, Elia Kaufmann, Rene Ranftl, Alexey Dosovitskiy, Vladlen Koltun, and Davide Scaramuzza. "Deep Drone Racing: From Simulation to Reality with Domain Randomization". In: *IEEE Transactions on Robotics (T-RO)* (2019).
  DOI: 10.1109/TRO.2019.2942989
  Links: PDF, Code, Video
- Daniele Palossi, **Antonio Loquercio**, Francesco Conti, Eric Flamand, Davide Scaramuzza, and Luca Benini. "A 64mW DNN-based Visual Navigation Engine for Autonomous Nano-Drones". In: *IEEE Internet of Things (IoT)* (2019).
  DOI: 10.1109/JIOT.2019.2917066
  Links: PDF, Code, Video

- **Antonio Loquercio**, A. Maqueda, C. del-Blanco, and D. Scaramuzza. "DroNet: Learning to Fly by Driving". In: *IEEE Robotics and Automation Letters (RA-L)* (2018).
  DOI: 10.1109/LRA.2018.2795643
  Links: PDF, Code, Video

Peer-Reviewed Conference Papers

The * symbol indicates shared first authorship.

- **Antonio Loquercio**, and Davide Scaramuzza. "Agile Autonomy: High-Speed Flight with On-Board Sensing and Computing". In: *Conference on Robotics and Intelligent Machines (I-RIM3D)* (2020).
  Links: PDF, Video Pitch
- Francesco Milano, **Antonio Loquercio**, Antoni Rosinol, Davide Scaramuzza, Luca Carlone. "Primal-Dual Mesh Convolutional Neural Networks". In: *Conference on Neural Information Processing Systems (NeurIPS)* (2020).
  Links: PDF, Code
- Yunlong Song, Selim Naji, Elia Kaufmann, **Antonio Loquercio**, and Davide Scaramuzza. "Flightmare:A Flexible Quadrotor Simulator". In: *Conference on Robotic Learning (CoRL)* (2020).
  Links: PDF, Video, Website
- Nico Messikommer, Daniel Gehrig, **Antonio Loquercio**, and Davide Scaramuzza. "Event-based Asynchronous Sparse Convolutional Networks". In: *IEEE European Conference on Computer Vision (ECCV)* (2020).
  DOI: 10.1007/978-3-030-58598-3_25
  Links: PDF, Video, Code
- Elia Kaufmann*, **Antonio Loquercio***, Rene Ranftl, Matthias Mueller, Vladlen Koltun, and Davide Scaramuzza. "Deep Drone Acrobatics". In: *Robotics: Science, and Systems (RSS)* (2020).
  DOI: 10.15607/RSS.2020.XVI.040
  Links: PDF, Video, Code
- Daniel Gehrig, **Antonio Loquercio**, Konstantinos G. Derpanis, and Davide Scaramuzza "End-to-End Learning of Representations for Asynchronous Event-Based Data". In: *IEEE International Conference on Computer Vision (ICCV)* (2019).
  DOI: 10.1109/ICCV.2019.00573
  Links: PDF, Video, Code
- Yanchao Yang*, **Antonio Loquercio***, Davide Scaramuzza, and Stefano Soatto. "Unsupervised Moving Object Detection via Contextual Information Separation". In: *IEEE Conference on Computer Vision and Pattern Recognition (CVPR)* (2018).
  DOI: 10.1109/CVPR.2019.00097
  Links: PDF, Video, Code

- Elia Kaufmann*, **Antonio Loquercio***, Rene Ranftl, Alexey Dosovitskiy, Vladlen Koltun, and Davide Scaramuzza. "Deep Drone Racing: Learning Agile Flight in Dynamic Environments". In: *Conference on Robotic Learning (CoRL)* (2018).
  Links: PDF, Code, Video
- Ana Maqueda, **Antonio Loquercio**, Guillermo Gallego, Narciso Garcia, and Davide Scaramuzza. "Event-based Vision meets Deep Learning on Steering Prediction for Self-driving Cars". In: *IEEE Conference on Computer Vision and Pattern Recognition (CVPR)* (2018).
  DOI: 10.1109/CVPR.2018.00568
  Links: PDF, Video
- Yawei Ye, Titus Cieslewski, **Antonio Loquercio**, and Davide Scaramuzza. "Place Recognition in Semi-Dense Maps: Geometric and Learning-Based Approaches". In: *British Machine Vision Conference (BMVC)* (2017).
  DOI: 10.5244/C.31.74
  Links: PDF, Poster

Open-Source Software

- DroNet: Learning to Fly by Driving
- Unsupervised Moving Object Detection via Contextual Information Separation
- Deep Drone Acrobatics
- Deep Drone Racing: From Simulation to Reality with Domain Randomization
- A General Framework for Uncertainty Estimation in Deep Learning

Awards

- **IEEE TRO King-Sun Fu Memorial Best Paper Award (Honorable Mention), 2020.**
- **IRIM-3D Best Paper Award (Honorable Mention), 2020.**
- **RSS Best Paper Award (Honorable Mention), 2020.**
- **CORL Best System Paper Award, 2018.**

Contents

Chapter 1
Introduction

This thesis presents algorithms for tightly-coupled robotic perception and action. According to this paradigm, in an iterative and infinite loop, action controls the amount of information coming from sensory data, and perception guides and provides feedback to action. Ecological psychology showed that the seamless integration of sensing and control is a fundamental feature of biological agents [55]. Can also artificial and embodied agents benefit from a tightly-coupled perception and action loop? My work addresses this question in the context of high-speed agile quadrotor flight.

In the following, we define as tightly-coupled any policy which directly predicts actions from (an history of) sensor observations, without explicit intermediate blocks.[1] Data-driven algorithms, *e.g.* neural networks, represent the ideal candidate to represent tightly-coupled policies: data-driven methods have recently achieved human-level performance in a series of standardized visual recognition and decision-making tasks. For example, the combination of supervised learning and large-scale datasets resulted in very high-performance systems for tasks like image classification [67, 71], segmentation [21], and detection [133]. Similarly, the combination of reinforcement learning algorithms with large-scale computing enabled the creation of systems that outperform humans in Atari games [113] and the complex game of Go [154]. However, all the previous tasks have a common denominator: they operate on a disembodied dataset or in controlled conditions, and they have an explicitly defined *score* function (a label is either wrong or correct in case of visual recognition, an action either decreases or increases the player's score in games). In general, such assumptions cannot be satisfied when a robot operates in unstructured and possibly dynamic environments. Therefore, while data-driven methods have the potential to

---

[1] Different research communities have associated different, and potentially ambiguous, meanings to the term tightly coupled. For instance, in state estimation [182], it refers to the problem of planning actions that maximize the visibility of landmarks during navigation.

© The Author(s), under exclusive license to Springer Nature Switzerland AG 2023
A. Loquercio, *Agile Autonomy: Learning High-Speed Vision-Based Flight*, Springer Tracts in Advanced Robotics 153, https://doi.org/10.1007/978-3-031-27288-2_1

Fig. 1.1 A tightly-coupled perception-action loop, powered by the combination of traditional model-based robotics and state-of-the-art machine learning tools, enables autonomous high-speed operation in unstructured environments. The quadrotor is required to reach a waypoint 50 m in front of it given by a user. The environment is not known in advance, and the robot has only access to on-board sensing and computation (no GPS or external tracking systems are used)

enhance robots' perception and control, their application to robotics poses a unique set of fundamental and technical challenges. Specifically:

- How can robots collect enough data to learn the complex mapping between noisy sensor measurements and motor skills?
- How can we trasfer knowledge between different domain or tasks?
- How can we verify and validate the safety of a deep sensorimotor policy controlling a robot?
- How can the interactive and embodied nature of robots be used to learn complex perception tasks?

To address these questions, my work combines theories and methods from machine learning, computer vision, and robotics, along with inspiration from cognitive science and psychology.

As a common demonstrator, I will consider the problem of high-speed quadrotor navigation in previously unknown real-wold environments (Fig. 1.1, hence in the wild) with only on-board sensing and computation (only a camera and an IMU are available to the agent). While previous work mainly considered manipulators or disembodied agents [3, 92, 128, 188, 189], quadrotors are the perfect test-bed to demonstrate the advantages of a tightly coupled perception and action loop. To date, only expert human pilots have been able to fully exploit their capabilities. Autonomous operation with onboard sensing and computation has been limited to low speeds. Indeed, agile quadrotor flight in unstructured environments requires low latency, robustness to perception disturbances, *e.g.* motion blur, and high precision, since the slack to avoid a collision is extremely limited. These characteristics push the boundaries of current state-of-the-art perception and control systems. My research shows that directly mapping sensory observations to navigation commands is a key principle to enable high-speed navigation. By mastering such a challenging navigation task, this work presents compelling evidence that the coupling of sensing and control is a fundamental step towards the development of a general-purpose robot autonomy in the physical world.

This thesis is split into three parts. First, given the difficulties to collect training data with quadrotors in the real world, I introduce an approach to transfer knowledge between domains, *e.g.* from simulation to reality. I also demonstrate the applicability of this method for a set of tasks, namely drone racing, acrobatic flight, and navigation of unstructured and dynamic environments. Second, I present a general algorithm to decode the uncertainty of neural network predictions, providing a way to integrate neural networks into safety-critical robotics systems as drones. Finally, I exploit the interactive nature of robots to train policies without any external or explicit supervision.

All parts address the driving research question of this work—*Why learning tightly-coupled perception and action policies in robotics?*—from a different perspective but with the same goal: building the next generation of autonomous robots. To facilitate training and development, I argue that future navigation systems will strongly rely on simulation. Therefore, effective methods are required to transfer knowledge between the simulated and real domain (Part I). In addition, for being successfully integrated into safety critical robotics systems and accepted by both governmental agencies and the general public, end-to-end policies need to automatically detect possible failure cases, *e.g.* due to noise in the data of lack of sufficient training data (Part II). Finally, to adapt to the conditions they observe during deployment and cannot be anticipated or modeled in simulation, robots will need to adapt their knowledge in the absence or with very limited supervisor signal. In such cases, self-supervised cues, *e.g.* visual and temporal consistency, can provide enough learning signal to train or tune even complex perception policies (Part III).

This thesis is structured in the form of a collection of papers. An introductory section that highlights the concepts and ideas behind the thesis is followed by self-contained publications in the appendix. Section 1.1 states and motivates the research objectives of this work. Section 1.2 places this research in the context of the related work. Chapter 2 summarizes the papers in the appendix and their connections with respect to each other. Finally, Chap. 3 provides future research directions.

1.1 Motivation

During the last decade, commercial drones have been disrupting industries ranging from agriculture to transport, security, infrastructure, entertainment, and search and rescue. In contrast to their grounded counterparts, flying robots can cover large distances in a very limited amount of time, which makes them an essential tool in missions where a fast response is crucial. Among commercial drones, quadrotors are the most agile: thanks to their agility they can navigate through complex structures, fly through damaged buildings, and reach remote locations inaccessible to other robots.

However, developing fully autonomous quadrotors that can approach or even outperform the agility of birds or human drone pilots in unknown and cluttered environments is very challenging and still unsolved. Changing lighting conditions, perceptual aliasing (i.e. strong similarity between different places), and motion blur, all common during agile flight in unstructured environments, are well-known and unsolved problems of state-of-the-art perception systems [32, 42, 108, 123]. In addition, the uncertainties coming from approximate models of the environment, sensors, and actuators, along with the requirements of closed-loop safety and a low latency perception-action loop add a significant layer of complexity to the problem of high-speed navigation.

Current state-of-the-art works addressed agile quadrotor navigation by splitting the task into a series of consecutive subtasks: perception, map building, planning, and control. This approach, illustrated in Fig. 1.2, constitutes the foundation of almost all current robotic systems, from autonomous cars to drones. The division of the navigation task into sequential subtasks is attractive from an engineering perspective since it enables parallel progress on each component and makes the overall system interpretable. However, it leads to pipelines that largely neglect interactions between the different stages and thus compound errors [182]. These errors are generated by sensor noise, *e.g.* motion blur or missing pixels, and have both a systematic and aleatoric nature. While the aleatoric part can only be accounted for in expectation, the systematic nature can be completely addressed. However, engineering robustness to all possible failure cases of a sensor is difficult to impossible. Therefore, noisy perception leads to imperfect mapping, suboptimal planning, and inaccurate control. In addition, the sequential nature of the blocks also introduces additional latency, which can be deterimental for agile flight [35]. While these issues can be mitigated to some degree by careful hand-tuning and engineering, the divide-and-conquer principle that has been prevalent in research on autonomous flight in unknown environments for

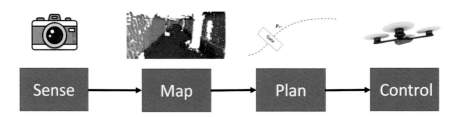

Fig. 1.2 The majority of existing methods for navigation in previously unknown environments divides the navigation problem into a series of consecutive subtasks. While this process is attractive from an engineering prospective and favours interpretability, it completely neglects the interactions between the different stages and thus compound errors

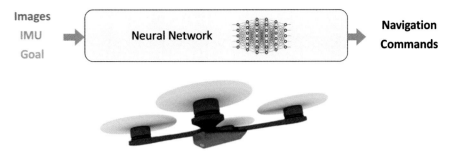

Fig. 1.3 Conversely to traditional methods, my thesis proposes a tight integration between sensing and control via a deep sensorimotor policy. This tight perception-action loop decreases the latency between perception and action and improves robustness against perception artifacts, *e.g.* motion blur or sensor noise. However, it also comes with several challenges, *i.e.* sample complexity, interpretability, and generalization to new environments and conditions, which this thesis aims to address

many years imposes fundamental limits on the speed and agility that a robotic system can achieve [99].

In contrast to these traditional pipelines, I propose a seamless and tight integration of perception and action through a deep sensorimotor policy (Fig. 1.3). This holistic approach comes with a series of advantages and challenges with respect to its traditional counterpart, which I describe in the next section.

1.1.1 Advantages

A tight perception-action loop, powered by advanced learning-based algorithms, has a series of advantages.

Low Latency. We define the *sensor-action latency* as the time required for a sensor measurement to influence the action executed by a robot. The action can be at a different level of abstraction, and it can vary from low-level motor commands to high-level velocity or position commands. However, the higher the action level, the longer it takes to transform it into motor commands. In the case of a deep sensorimotor policy, this latency generally corresponds to the time of a neural network forward pass. Thanks to the continuously evolving technology on hardware acceleration for deep networks, the latency of a sensorimotor policy is now in the order of milliseconds on on-board hardware, even for large architectures, and up to 10 times lower than the one of a sequential system [104]. Having a low latency enables a fast control loop, which is fundamental, for instance, to promptly react to unexpected obstacles. In addition, the combination of deep sensorimotor policies and hardware acceleration enables closed-loop control of nanorobots with extremely low power and computational budget [126].

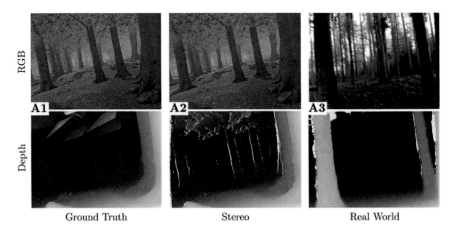

Ground Truth   Stereo   Real World

Fig. 1.4 The non-idealities of sensor observations can strongly affect performance in traditional pipelines since those errors compound over the pipeline. In contrast, a tight perception-action cycle learns to cope with the systematic nature of those errors by leveraging regularities in the data. Learning those regularities does not necessarily require training in the real world: for instance, a ground-truth depth image (A1) is very different from a depth image collected in the real world (A3). However, a simulated depth constructed by stereo matching (A2) contains the typical failure cases of a real depth sensor, e.g. missing values and noise

Robustness to Imperfect Perception. All sensors have nonidealities. For example, a depth sensor operating on stereo images produces observations with discretization artifacts, missing data, or erroneous measurements (c.f. Fig. 1.4). Similarly, an image recorded from a camera can be corrupted by motion-blur or artifacts caused by the low dynamic range or the image capturing method (*e.g.* global or local shutter). When provided with enough training data, deep sensorimotor policies become robust to perception nonidealities. Indeed, such nonidealities generally have an epistemic, *i.e.* systematic, nature, which a neural network learns to leverage by using regularities in the data. Directly identifying systematic errors and engineering solutions to them is a very challenging task. For this reason, traditional methods are generally more prone to failures due to compounding errors through the pipeline.

Ease to Develop and Deploy. The fast-growing scale of machine learning research pushed the development of many open-source tools for the creation and validation of learning-based systems. These tools are not only easy to use on normal computers but are sometimes even optimized for computationally constrained on-board hardware [126]. In addition, the machine learning community has strongly promoted an open-source philosophy aimed at reproducibility, and it spent large efforts to build simulators [29, 45, 150, 156] and datasets [25, 28, 96], which are all readily available to the machine learning practitioner. Conversely, traditional methods are significantly more diverse and specialized than their learning counterparts and often rely on the end-user to both develop the low-level functionalities and optimize them on hardware. Despite this limitation being not fundamental (large companies have hundreds of experienced engineers to develop their systems), it strongly affects the speed at which new ideas can be implemented and tested on a physical system.

1.1.2 Challenges

Despite their numerous advantages over traditional sequential methods, learning-based systems present several challenges that need to be overcome to unlock their full potential for computer vision and robotics. The main challenge arises from their data complexity: a sufficiently large and representative dataset is required to successfully train a deep sensorimotor policy for a specific task. While domain knowledge can help to decrease the data complexity, it is not necessarily clear how to inject such prior knowledge into the training process. In addition, contrary to traditional methods, which are transparent and easy to interpret, learning-based systems are often black-boxes and therefore difficult to interpret and debug. Thus, guaranteeing performance during deployment is challenging—especially for algorithms that are expected to cope with the inherent uncertainties of the real world.

Sample Complexity. Machine learning algorithms are data-hungry. Popular datasets for training networks on standard visual recognition tasks generally have training samples in the order of millions. When the task is closed-loop control of a robot the data complexity can be even larger, since it is necessary to expose the system to all the possible situations it could encounter during operation (*i.e.* to sufficiently explore the state space). Building such training datasets is a tedious and expensive process, but also raises the problem of how to collect enough "negative" experiences (*e.g.* very close to obstacles). This region of the state space is generally far from the positive (*i.e.* safe) states observed by an expert (*e.g.* a human pilot) and challenging to observe and sufficiently represent in a dataset. Addressing this problem is one of the core endeavors of my thesis. I propose to tackle this problem by recycling data collected from different platforms [106] or by training the sensorimotor policies entirely in simulation, as shown in Fig. 1.5 (Papers [81, 104, 105]). Doing so waives the requirement of a human expert to provide demonstrations, it cannot harm the physical system during training, and it can be used to learn behaviors that are challenging even for the best human pilots. In Sect. 2.1, I show how it is possible to enable knowledge transfer between different platforms, *e.g.* a car and a drone, or different domains, *e.g.* simulation and the real world.

Use of Domain Knowledge. Over the years, roboticists have developed a large knowledge base in how to model [98] and control [12] complex systems with high accuracy. Using this knowledge to decrease the data complexity and to favor generalization to new environments and tasks is a major opportunity for learning-based methods. However, how can this knowledge be injected into a deep sensorimotor policy? In this thesis, I propose two different ways to address this problem: (i) use domain knowledge to build experts, generally operating on *privileged* information, that can

Train   **Test**

Fig. 1.5 Learning the complex mapping between noisy sensor observations and actions generally requires large and diverse training datasets. Collecting these datasets in the real world is a tedious and expensive procedure, that also raises fundamental questions on how to sufficiently cover the state space to enable closed-loop performance. To address this problem, I propose to train sensorimotor policies entirely in simulation and then transfer the acquired knowledge to a physical platform via domain-invariant abstraction

supervise the training of the sensorimotor policy [81, 104, 105], and (ii) to make the sensorimotor student operate on high-level control commands, *e.g.* a desired trajectory, which is then executed by a low-level model-based controller [104–106]. Both approaches simplify the training procedure with respect to model-free approaches, in which the laws of physics need to be learned from scratch.

Interpretability. When operating a physical robot, wrong decisions could not only fail the mission but also put human lives at risk, *e.g.* if the robot is an autonomous car or a medical device. Traditional methods, generally composed of sequential modules tackling different sub-tasks, can naturally be interpreted, tested, and inspected for the identification of errors. In contrast, deep sensorimotor policies, directly mapping noisy sensor observations to actions, do not offer a straightforward way to be interpreted or validated. In addition, neural network predictions are known to be unreliable when the inputs are out of the training distribution [47]. To tackle this problem, in Sect. 2.2 I present a general framework for uncertainty estimation in deep neural network predictions. Having a measure of uncertainty enables the detection of sensor failure, out-of-distribution samples, and gives valuable feedback during closed-loop control [103].

Generalization. To build a general-purpose robot autonomy, artificial agents need to develop skills to quickly adapt to new environments and conditions, while efficiently learning new tasks in a handful of trials. However, current robot automation systems, whether or not data-driven, suffer from a limited generalization. Indeed, simply changing the operation environment, but keeping the task definition unchanged, might require intensive re-training or fine-tuning of the system. To improve robustness to visual changes in the environment, I have proposed to use domain adaption [104, 105].

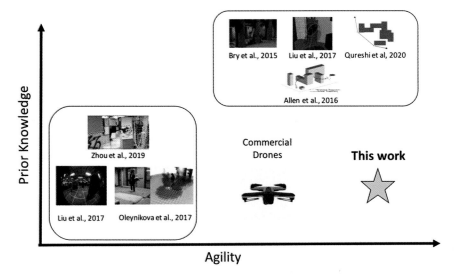

Fig. 1.6 Overview of existing approaches for drone navigation in challenging and cluttered environments. While state-of-the-art approaches make strong assumptions in terms of prior knowledge or are only limited to low speeds, the primary goal of this thesis is to develop methods that can exploit the agility of quadrotors with no prior knowledge about the environment or external sensing

In addition, to enable adaption, I have developed techniques to train from unsupervised or weakly-supervised data, which can be collected during operation and directly leverage the interactive nature of robots [102, 180]. However, generalization remains one of the open challenges of robotics and, in my opinion, holds the key to a new era of general-purpose robot autonomy in the real world (Fig. 1.6).

1.2 Related Work

This section summarizes the state of the art in perception and control for agile drone navigation. It also indicates how the work of this thesis relates to and improves on state-of-the-art methods.

1.2.1 Autonomous Drone Navigation in Unknown Environments

Autonomous, vision-based drone navigation gives rise to fundamental challenges for both perception and control. On the perception side, motion blur, challenging lighting

Fig. 1.7 While prior works can exploit the ability of flying machines to perform agile maneuvers, they assume that near-perfect state estimation is available during the maneuver. In practice, this entails disregarding the challenges of perception and instrumenting the environment with dedicated sensors

conditions, and aliasing can cause severe drift in vision-based state estimation [42, 108, 119]. Other sensory modalities, *e.g.* LIDAR or event-based cameras, could partially alleviate these problems [17, 137]. Those sensors are however either too bulky or too expensive to be used on small quadrotors. Moreover, state-of-the-art state estimation methods are designed for a predominantly-static world, where no dynamic changes to the environment occur.

From the control perspective, plenty of work has been done to enable high-speed navigation, both in the context of autonomous drones (Fig. 1.7) [1, 107, 115, 116] and autonomous cars [79, 83, 89, 142]. Lupashin et al. [107] propose iterative learning of control strategies to enable platforms to perform multiple flips. Abbeel et al. [1] learn to perform a series of acrobatic maneuvers with autonomous helicopters. Their algorithm leverages expert pilot demonstrations to learn task-specific controllers. Similarly, Li et al. [94] propose an imitation learning approach for training visuo-motor agents for the task of quadrotor flight. While these works proved the ability to fly machines to perform agile maneuvers, they did not consider the perception problem. Indeed, they assume that near-perfect state estimation is available during the maneuver, which in practice requires instrumenting the environment with dedicated sensors.

Agile flight with on-board sensing (camera and IMU) and computation is a challenging research problem, especially for navigation in previously unknown real-world environments. The first attempts in this direction were made by Shen et al. [152], who demonstrated agile vision-based flight. The work was limited to low-acceleration trajectories, therefore only accounting for part of the control and perception problems encountered at high speed. More recently, Loianno et al. [100] and Falanga et al. [37] demonstrated aggressive flight through narrow gaps with only onboard sensing. Even though these maneuvers are agile, they are very short and cannot be repeated without re-initializing the estimation pipeline. Overall, these works are specifically tailored to the operation environment, *e.g.* a narrow opening [37, 100], and cannot be used for general navigation of previously unknown environments.

We will now discuss approaches that have been proposed to overcome the aforementioned problems. A taxonomy of prior works is presented in Fig. 1.6.

Traditional Methods. Various approaches to enable autonomous flight have been
proposed in the literature. Some works tackle only perception and build high-quality
maps from imperfect measurements [10, 34, 46, 68, 147], while others focus on
planning without considering perception errors [4, 18, 97, 135]. Numerous systems
that combine online mapping with traditional planning algorithms have been proposed to achieve autonomous flight in previously unknown environments [8, 24,
114, 123, 124, 141, 169, 182, 184].

Oleynikova et al. [123] propose a computationally efficient method–Voxblox–to
build a map of the environment with associated obstacle distance information. This
mapping scheme is combined with a trajectory optimization method to navigate a
drone in a cluttered environment with only on-board computation. However, this
approach is only demonstrated at relatively low speeds ($1 m s^1$) and in combination
with off-board sensing (near-perfect state estimation coming from a Vicon tracking
module), which restricts its applications to controlled environments (Fig. 1.8).

In the context of the DARPA Fast Lightweight Autonomy (FLA) program,
Mohta et al. [114] developed a system for fast and robust aerial robot autonomous
navigation in cluttered, GPS-denied environments (Fig. 1.9). The system assumes
little to no prior knowledge of the scene and relies on on-board sensing and computation for state estimation, control, mapping, and planning. Navigation commands
are predicted by a system by combining a lidar-based mapping module and an A^*
planner module, and they are executed with a cascade of LQR controllers. However,
the system requires bulky sensors, *e.g.* Lidar, it is very computationally expensive
and prone to compounding errors over the pipeline. Such requirements pose a limit
on the agility of the drone, preventing high-speed agile navigation.

Zhou et al. [184] introduced a robust and perception-aware replanning framework
to support fast and safe flight, which currently constitutes the state-of-the-art for

Fig. 1.8 A quadrotor navigating to a point behind the mannequin by combining a voxblox-generated
map with a local re-planning trajectory optimization-based method. All computation runs entirely
on-board, but state sensing is off-board and generated by an external tracking module

(a)   (b)

Fig. 1.9 The system proposed by Mohta et al. can navigate previously unknown environments with only on-board sensing and computation **a**. However, it relies on a bulky computational and sensing unit **b**, which limits the agility and speed of the system

(a)   (b)

Fig. 1.10 The system proposed by Zhou et al. enables agile flight in indoor **a** and outdoor **b** environments. However, it is limited to static environments and relatively low speeds. In addition, its sequential nature makes it subject to sensing errors, typical during agile flight, which compound over the pipeline

vision-based agile navigation (Fig. 1.10). Its key contributions are a path-guided optimization approach to efficiently find feasible and high-quality trajectories and a perception and risk-aware planning strategy to actively observe and avoid unknown obstacles. This approach demonstrated agile drone navigation in previously unknown cluttered environments with only on-board vision-based perception and computation. However, this approach is still limited to static environments, and it only achieves maximum speeds of 3 m s^1, thus not exploiting the full agility of drones.

Overall, tfhe division of the navigation task into the mapping and planning sub-tasks is attractive from an engineering perspective, since it enables parallel progress on each component and makes the overall system interpretable. However, it leads to pipelines that largely neglect interactions between the different stages and thus compound errors [182]. Their sequential nature also introduces additional latency, making high-speed and agile maneuvers difficult to impossible [35]. While these issues can be mitigated to some degree by careful hand-tuning and engineering, the divide-and-conquer principle that has been prevalent in research on autonomous flight in unknown environments for many years imposes fundamental limits on the speed and agility that a robotic system can achieve [99].

Fig. 1.11 To overcome the limitations of traditional navigation pipelines, previous work proposed to learn complete navigation policies directly from data, collected either by a human (left) [138] or an automated procedure (right) [50]. However, to address the policies' sample complexity, they imposed strong motion-constraints, *e.g.* planar motion, which do not exploit the agility of the platform

Data-Driven Methods. In contrast to these traditional pipelines, some recent works propose to learn navigation policies directly from data, abolishing the explicit division between mapping and planning (Fig. 1.11). These policies are trained by imitating a human [138] or an algorithmic expert [181], or directly from experience collected in the real world [50]. Zhang et al. [181] train a deep sensorimotor policy directly from the demonstrations provided by an MPC controller. While the latter has access to the full state of the platform and knowledge of obstacle positions, the policy only observes on-board sensor measurements, *i.e.* laser range finder, and inertial measurements. Gandhi et al. [50] predict the collision probability with a neural network trained with real-world data and command the drone towards the direction minimizing the collision risk. As the number of samples required to train general navigation policies is very high, existing approaches impose constraints on the quadrotor's motion model, for example by constraining the platform to planar motion [50, 138, 181] and/or discrete actions [50, 138], at the cost of reduced maneuverability and agility.

Transfer from Simulation to Reality. Learning navigation policies from real data has a shortcoming: the high cost of generating training data in the physical world. Data needs to be carefully collected and annotated, which can involve significant time and resources. To address this problem, a recent line of work has investigated the possibility of training a policy in simulation and then deploying it on a real system (Fig. 1.12). Work on the transfer of sensorimotor control policies has mainly dealt with manual grasping and manipulation [13, 61, 74, 140, 144, 178]. In the context of drone control, Sadeghi and Levine [143] learned a collision-avoidance policy by using 3D simulation with extensive domain randomization [165]. Despite this policy being able to control a drone in the real world, the proposed model-free training procedure requires constraining the actuation space, *i.e.* discrete actions, to

Fig. 1.12 Instead of training with real-world data, prior work explored the possibility to train entirely in simulation and then transfer to the real platform using abstractions of inputs (left) [117] or domain randomization (right) [143]. Yet, these transfer procedures require careful tuning of abstraction/randomization parameters and constraint the platforms to relatively slow motion in 2D

converge. Indeed, the policy only achieves very low speeds and cannot profit from the agility of the drone.

In driving scenarios, synthetic data was mainly used to train perception systems for high-level tasks, such as semantic segmentation and object detection [76, 136]. One exception is the work of Müller et al. [117], who propose a novel strategy based on abstraction to transfer a policy trained in simulation on a physical ground vehicle. Specifically, they propose to train a control policy from abstract representations, *e.g.* semantic segmentation, instead of the raw sensory inputs, *e.g.* color images. This training strategy strongly simplifies the transfer problem, since the abstract representations are more similar between domains than the raw sensor measurements. Indeed, they empirically show that their policy trained in simulation generalizes to multiple environments and lighting conditions.

In Paper [81], I derive the theoretical foundations of simulation to reality via abstraction using information theory [2]. The resulting framework is then validated on multiple tasks including drone racing [105], acrobatic flight [81], and high-speed flight in the wild [104].

Simulators. The landscape of currently available simulators for quadrotor control is fragmented (Fig. 1.13): some are extremely fast, *e.g.* Mujoco [166], while others have either really accurate dynamics [45, 72] or highly photo-realistic rendering [60]. Both RotorS [45] and Hector [88] are popular Micro Aerial Vehicle (MAV) simulators built on Gazebo [87], which is a general robotic simulation platform and generally used with the popular Robot Operating System (ROS). These gazebo-based simulators have the capability of accessing multiple high-performance physics engines and simulating various sensors, ranging from laser range finders to RGB cameras. Nevertheless, Gazebo has limited rendering capabilities and is not designed for efficient parallel dynamics simulation, which makes it unattractive for the development of learning-based systems. AirSim [150] is an open-source photo-realistic simulator

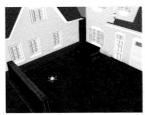

Fig. 1.13 Several quadrotor simulators are available to develop, train, and test navigation algorithms. While some offer very accurate dynamics (left) [45], others have highly photo-realistic rendering (center) [150]. I have contributed to build Flightmare (right) [156], a simulator which offers both and is specifically tailored to machine learning applications

for drones built on Unreal Engine. Despite being very photorealistic, AirSim has tight integration between the physics and the rendering module. This is a disadvantage for two reasons: (i) the physics engine is not specialized for racing quadrotors and cannot be easily modified, and (ii) it strongly limits the simulation speed if images are not required. This limitation makes it difficult to apply the simulator to challenging model-free reinforcement learning tasks. FlightGoggles [60] is a photo-realistic sensor simulator for perception-driven robotic vehicles. This simulator is very useful for rendering camera images given trajectories and inertial measurements from flying vehicles in real-world, in which the collected dataset [6] is used for testing vision-based algorithms. However, it has a very limited sensor suite and can only simulate a single quadrotor at a time, limiting the speed at which large-scale datasets can be collected. Eventually, Muller et al. [118] built Sim4CV, a photo-realistic quadrotor simulator that is however only aimed at computer vision applications.

During my Ph.D., I have contributed to building a simulator that is specifically tailored to machine learning applications: Flightmare [156]. Flightmare is composed of two main components: a configurable rendering engine built on Unity and a flexible physics engine for dynamics simulation. Those two components are decoupled and can run independently of each other. This makes our simulator extremely fast: rendering achieves speeds of up to 230 Hz, while physics simulation of up to 200,000 Hz on a laptop. I have used this simulator for the development of [104]. Specifically, I have used Flightmare to train a policy that can navigate quadrotors at high speed in cluttered environments. The policy, trained entirely in simulation, was then deployed in multiple indoor and outdoor scenarios in the physical world.

1.2.2   Autonomous Drone Racing

The popularity of drone racing has recently kindled significant interest in the robotics research community (Fig. 1.14). The classic solution to this problem is image-based visual servoing, where a robot is given a set of target locations in the form of reference images or patterns. Target locations are then identified and tracked with hand-crafted

Fig. 1.14 The classic approach to autonomous drone racing is to detect the next gate and steer the drone towards it (left) [77]. To account for the cases when no gate is visible, it is possible to combine state estimation and gate detection through Kalman filtering (right) [41]. Multi-agent autonomous drone racing also raises interesting game-theoretical problems (center) [157]

detectors [37, 93, 163]. Li et al. [95] tailor this approach to the problem of drone racing. Specifically, they propose to extract gate information from images and then fuse it with attitude estimates from an inertial measurement unit (IMU) to guide the drone towards the next visible gate. While the approach is computationally very light-weight, it struggles with scenarios where multiple gates are visible and does not allow to employ more sophisticated planning and control algorithms which, *e.g.*, plan several gates ahead. Jung et al. [77] retrieve a bounding box of the gate and a line-of-sight-based control law aided by optic flow is then used to steer the drone towards the detected gate. However, these methods cannot cope with the cases when no gate is visible, which is typical during drone racing.

To account for this problem, previous work [80] combines the gate detector with a pre-computed global plan and a visual odometry system with uncertainty estimation and Kalman filtering. Foehn et al. [41] extend this idea to account for multiple visible gates and improve the localization and planning scheme. Overall, the above methods strongly rely on the quality of the gate detection system, which can however quickly become unreliable in the presence of occlusions, partial visibility, and motion blur. In addition, the separation between perception, planning, and control may lead to compounding errors over the pipeline, leading to a lack of robustness to perception disturbances, *e.g.* challenging illumination conditions. In Paper [105], I show that combining the perception and planning block into a convolutional neural network and leveraging the abundance of simulated data, favors robustness against dynamic environments and perception disturbances.

1.2.3 Uncertainty Estimation for Safe Deep Learning

To successfully integrate learning methods into robotics, neural networks should reliably estimate the uncertainty in their predictions [162]. In this section, I review existing methods for uncertainty estimation and some of their applications to robotics.

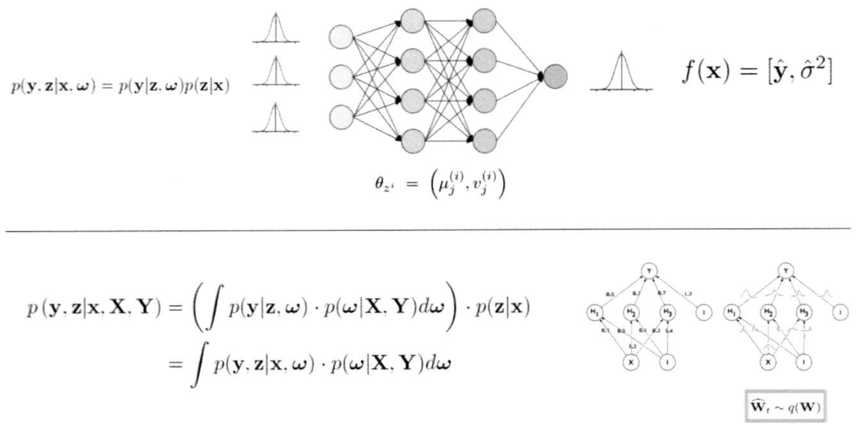

$$p(\mathbf{y}, \mathbf{z}|\mathbf{x}, \boldsymbol{\omega}) = p(\mathbf{y}|\mathbf{z}, \boldsymbol{\omega})p(\mathbf{z}|\mathbf{x})$$

$$f(\mathbf{x}) = [\hat{\mathbf{y}}, \hat{\sigma}^2]$$

$$\theta_{z^i} = \left(\mu_j^{(i)}, v_j^{(i)}\right)$$

$$p(\mathbf{y}, \mathbf{z}|\mathbf{x}, \mathbf{X}, \mathbf{Y}) = \left(\int p(\mathbf{y}|\mathbf{z}, \boldsymbol{\omega}) \cdot p(\boldsymbol{\omega}|\mathbf{X}, \mathbf{Y})d\boldsymbol{\omega}\right) \cdot p(\mathbf{z}|\mathbf{x})$$

$$= \int p(\mathbf{y}, \mathbf{z}|\mathbf{x}, \boldsymbol{\omega}) \cdot p(\boldsymbol{\omega}|\mathbf{X}, \mathbf{Y})d\boldsymbol{\omega}$$

$$\widehat{\mathbf{W}}_t \sim q(\mathbf{W})$$

Fig. 1.15 Prediction uncertainty in deep neural networks generally derives from two sources: data uncertainty and model uncertainty. Data uncertainty (above) arises because of noise in the data, usually caused by the sensors' non-idealities. Such uncertainty can be estimated through Bayesian Belief Networks [52] or specialized training [84]. Model uncertainty (below) is generated from unbalances in the training data distribution, and it is generally estimated through sampling [47]

Estimating Uncertainties in Neural Networks Predictions. A neural network is generally composed of a large number of parameters and non-linear activation functions, which makes the (multi-modal) posterior distribution of network predictions intractable. To approximate the posterior, existing methods deploy different techniques, mainly based on Bayesian inference and Monte-Carlo sampling (Fig. 1.15).

To recover probabilistic predictions, Bayesian approaches represent neural network weights through parametric distributions, *e.g.*, exponential-family [11, 44, 69, 173]. Consequently, networks' predictions can also be represented by the same distributions and can be analytically computed using non-linear belief networks [44] or graphical models [159]. More recently, Wang et al. [173] propose natural parameter networks, that model inputs, parameters, nodes, and targets by Gaussian distributions. Overall, this family of approaches can recover uncertainties in a principled way. However, they generally increase the number of trainable parameters in a super-linear fashion and require specific optimization techniques [69] which limits their impact in real-world applications.

To decrease the computational burden, Gast et al. [52] proposed to replace the network's input, activations, and outputs with distributions, while keeping the network's weights deterministic. Similarly, probabilistic deep state-space models retrieve data uncertainty in sequential data and use it for learning-based filtering [43, 177]. However, disregarding weights uncertainty generally results in over-confident predictions, in particular for inputs not well represented in the training data.

Instead of representing neural network parameters and activations by probability distributions, another class of methods proposed to use Monte-Carlo (MC) sampling to estimate uncertainty. The MC samples are generally computed using an ensemble of neural networks. The prediction ensemble could either be generated by differently trained networks [78, 85, 90], or by keeping drop-out at test-time [47]. While this class of approaches can represent well the multi-modal posterior by sampling, it cannot generally represent data uncertainty, due for example to sensor noise. A possible solution is to tune the dropout rates [48], however, it is always possible to construct examples where this approach would generate erroneous predictions [73]. Recent work shows the possibility to estimate model uncertainty without sampling, decreasing the run-time computational budget required for uncertainty estimation [130, 151] However, also these methods do specifically account for the data uncertainty.

To model data uncertainty, Kendall et al. [84] proposed to add to each output a "variance" variable, which is trained by a maximum-likelihood (a.k.a. heteroscedastic) loss on data. Combined with Monte-Carlo sampling, this approach can predict both the model and data uncertainty. However, this method requires changing the architecture, due to the variance addition, and using the heteroscedastic loss for training, which is not always an option for a general task.

Akin to many of the aforementioned methods, in Paper [103] I use Monte-Carlo samples to predict model uncertainty. Through several experiments, I show why this type of uncertainty, generally ignored or loosely modeled by Bayesian methods [52], cannot be disregarded. In addition to Monte-Carlo sampling, my approach also computes the prediction uncertainty due to the sensors' noise by using gaussian belief networks [44] and assumed density filtering [14]. Therefore, my approach can recover the full prediction uncertainty for any given (and possibly already trained) neural network, without requiring any architectural or optimization change.

Uncertainty Estimation in Robotics. Given the paramount importance of safety, autonomous driving research has allocated a lot of attention to the problem of uncertainty estimation, from both the perception [38, 120] and the control side [78, 85] (Fig. 1.16). Feng. et al. [38] showed an increase in performance and reliability of a 3D Lidar vehicle detection system by adding uncertainty estimates to the detection pipeline. Predicting uncertainty was also shown to be fundamental to cope with sensor failures in autonomous driving [85], and to speed-up the reinforcement learning process on a real robot [78].

For the task of autonomous drone racing, Kaufmann et al. [80] demonstrated the possibility to combine optimal control methods to a network-based perception system by using uncertainty estimation and filtering. Also for the task of robot manipulation, uncertainty estimation was shown to play a fundamental role to increase the learning efficiency and guarantee the manipulator safety [23, 158].

In Paper [103], I show that uncertainty estimation is beneficial in both perception and control tasks. In perception tasks, I show that neural networks naturally capture the uncertainty of inherently ambiguous tasks, *e.g.* future motion prediction. In the task of closed-loop control of a quadrotor, I demonstrate that identifying possible wrong predictions or out-of-distribution input samples is beneficial to control performance.

Fig. 1.16 Estimating uncertainty is of paramount importance to integrate learning-based methods into robotics systems. Kahn et al. [78] used model-uncertainty to speed-up the learning process for an autonomous car (left). Kaufmann et al. [80] estimated the uncertainty of gate detection to enable bayesian decision making during drone racing (center). Chua et al. [23] accounted for model errors during model-based reinforcement learning by leveraging their uncertainty (left)

1.2.4   Unsupervised and Weakly-Supervised Learning Robot Vision

Robust perceptions are necessary for enabling robots to understand and interact with their surroundings. The majority of state-of-the-art perception algorithms, included some of the methods presented in this thesis, rely on large annotated datasets or specifically designed reward functions. In contrast, natural agents learn (part of) their skills with interaction with the environment. This section explores prior work that tries to invert this paradigm and learn in a more natural and scalable fashion. Specifically, I examine this problem on a low-level task (*i.e.* depth estimation) and a high-level task (*i.e.* moving object detection).

3D Perception. The problem of recovering the three-dimensional structure of a scene from its two-dimensional projections has been long studied in computer vision [65, 66, 101, 170]. Classic methods are based on multi-view projective geometry [64]. However, key characteristics of these classic methods are that they crucially rely on projective geometry, require laborious hand-engineering, and are not able to exploit non-motion-related depth cues.

To make optimal use of all depth cues, machine learning methods can either be integrated into the classic pipeline or replace it altogether (Fig. 1.17). The challenge for supervised learning methods is the collection of training data: obtaining ground truth camera poses and geometry for large realistic scenes can be extremely challenging. Therefore, while supervised learning methods have demonstrated impressive results [30, 131, 171, 186], it is desirable to develop algorithms that function in the absence of large annotated datasets.

Unsupervised (or self-supervised) learning provides an attractive alternative to label-hungry supervised learning. The dominant approach is inspired by classic 3D reconstruction techniques and makes use of projective geometry and photometric consistency across frames. Among the methods for learning depth maps, some operate in the stereo setup [51, 56], while others address the more challenging monocular

Fig. 1.17 To recover the 3D structure of the scene, state-of-the-art methods train neural networks with large annotated datasets (left) [171], which are expensive and tedious to collect. Self-supervised methods wave the requirement for training datasets by using projective geometry and photometric consistency across frames (right) [57]. However, these methods suffer when the views contain reflective surfaces or are poorly textured

setup, where the training data consists of monocular videos with arbitrary camera motions between the frames [109, 187]. Reprojection-based approaches can often yield good results in driving scenarios, but they crucially rely on geometric equations and precisely known camera parameters (one notable exception being the recent work in [57], which learns the camera parameters automatically) and enough textured views.

In Paper [102], I show how is possible to train a neural network to predict metric depth with the feedback an agent would observe interacting with the environment: a sequence of RGB frames and very sparse depth measurement (down to one pixel per image). In contrast to geometric unsupervised methods, my approach does not require knowing the camera parameters in advance and is robust in low-textured indoor scenarios.

Several works similar to mine aim to learn 3D representations without explicitly applying geometric equations [33, 134, 164]. However, they build implicit 3D representations, and therefore they cannot be directly used for downstream robotic tasks such as navigation or motion planning. Moreover, at training time it requires knowing the camera pose associated with each image. My method, in contrast, does not require camera poses and grounds its predictions in the physical world via very sparse depth supervision. This allows us to learn an explicit 3D representation in the form of depth maps.

Moving Object Detection. The ability to detect moving objects is primal for animals and robots alike, so there is a long history of motion-based segmentation, or moving object detection (Fig. 1.18). Early attempts to explicitly model occlusions include the layer model [174] with piecewise affine regions, with computational complexity improvements using graph-based methods [153] and variational inference [16, 26, 160, 179] to jointly optimize for motion estimation and segmentation. Ochs et al. [121] use long-term temporal consistency and color constancy, making however the optimization more difficult and sensitive to parameter choices. Similar ideas

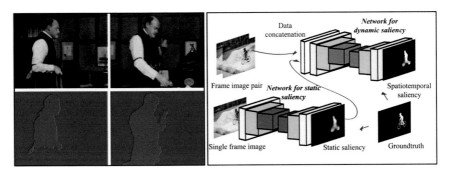

Fig. 1.18 Classical methods for moving object detection relies on super-pixels and color constancy to assign object relationship over time (left) [121]. However, they are generally very sensitive to hyper-parameters or require the solution of differential equations at run time. Conversely, data-driven methods have a very good generalization and outperform traditional ones on public benchmarks (right) [167]. However, they rely on large annotated datasets for training

were applied to motion detection in crowds [15], traffic monitoring [9] and medical image analysis [31].

More recent data-driven methods [22, 155, 167, 168] learn discriminative spatio-temporal features and differ mainly for the type of inputs and architectures. Inputs can be either image pairs [22, 155] or image plus dense optical flow [167, 168]. Architectures can be either time-independent [167], or with recurrent memory [155, 168]. Overall, those methods outperform traditional ones on benchmark datasets [121, 129], but at the cost of requiring a large amount of labeled training data and with evidence of poor generalization to previously unseen data.

In Paper [180], I present an approach for moving object detection that, similarly to the classic object detection literature, does not need any annotated training data. However, like modern learning methods, my approach can use contextual cues to help prediction, which would be impossible to engineer given the complexity of image formation and scene dynamics.

Chapter 2
Contributions

This chapter summarizes the key contributions of the papers that are reprinted in the appendix. It further highlights the connections between the individual results and refers to related work and video contributions.

In total, this research has been published in 4 peer-reviewed conference publications and 5 journal publications (one in the *IEEE Transactions on Robotics*, one in *Science Robotics*, and three in the *Robotics Automation Letters (RA-L)*).

These works led to several research awards and open-source software.

Awards.

- IEEE Transactions on Robotics King-Sun Fu Memorial Best Paper Award (Honorable Mention), 2020
- IRIM-3D Best Paper Award (Honorable Mention), 2020
- RSS Best Paper Award (Honorable Mention), 2020
- CORL Best System Paper Award, 2018

Software.

- Dronet: Learning to Fly by Driving
- Unsupervised Moving Object Detection via Contextual Information Separation
- Deep Drone Racing: From Simulation to Reality with Domain Randomization
- A General Framework for Uncertainty Estimation in Deep Learning
- Deep Drone Acrobatics

2.1 Transfer Learning for Agile Drone Navigation

In this part of the thesis, we consider the problem of short-range navigation for autonomous drones. In this task, the drone has access to only on-board sensor observations from a camera and an IMU, as well as a user-defined reference, indicating the goal or the desired path. This reference is not necessarily collision-free and it generally the output of an higher level planning or exploration algorithm [24]. The task consists of following this reference while possibly avoiding obstacles. Neither

© The Author(s), under exclusive license to Springer Nature Switzerland AG 2023
A. Loquercio, *Agile Autonomy: Learning High-Speed Vision-Based Flight*, Springer Tracts in Advanced Robotics 153, https://doi.org/10.1007/978-3-031-27288-2_2

prior knowledge about the application environment nor external information, *e.g.* GPS, are available to the agent.

One approach to train neural networks to address this task is to collect a set of human demonstrations and learn a policy from them [1]. However, this approach has some limitations. The most obvious one is related to the cost (both in terms of resources and time) to collect large enough datasets: expert human pilots are a scarce and expensive resource. In addition, data coming from human experts are more saddle than one might expect. How can we collect enough "negative" experiences, far from the positive (*i.e.* safe) states observed by a human pilot? How can we push the neural network to learn to navigate, and not just overfit to the particular scene/situation in which the data was collected?

The work conducted during this thesis contributes to addressing these questions in several ways. First, I developed an approach to train drones to fly in an urban environment by only collecting data from completely different platforms, *e.g.* grounded vehicles [106]. This approach was also applied for the control of palm-sized drones [126] in a collaborative effort between me and the Integrated System Lab (ISL) at ETH Zurich. Then, I developed a theoretically sound approach to transfer knowledge between platforms and domains, *e.g.* from simulation to the physical world. Such framework was used to train navigation algorithms for drone racing [105], acrobatics [81], and navigation in the wild [104]. These algorithms were trained exclusively in simulation and never saw a real image before being deployed on physical drones. I also contributed to building the infrastructure required to develop such algorithms by developing a high-quality drone simulator [156]. This simulator was openly released to the public to foster progress in the community.

2.1.1   Dronet: Learning to Fly by Driving

(P1) Antonio Loquercio, Ana I. Maqueda, Carlos R. del-Blanco, and Davide Scara-
      muzza. "Dronet: Learning to Fly by Driving". In: *IEEE Robotics Autom. Lett.*
      3.2 (2018), pp. 1088–1095

In unstructured and highly dynamic scenarios drones face numerous challenges to navigate autonomously feasibly and safely. In this work, we explore a data-driven approach to cope with the challenges encountered when flying in urban environments. Specifically, we propose DroNet, a convolutional neural network that can safely drive a drone through the streets of a city (Fig. 2.1). Designed as a fast 8-layers residual network, DroNet produces, for each single input image, two outputs: a steering angle, to keep the drone navigating while avoiding obstacles, and a collision probability to let the UAV recognize dangerous situations and promptly react to them. We train this policy from data collected by cars and bicycles, which, already integrated into urban environments, offer the perfect opportunity to collect data in a scalable fashion. This work demonstrates that specifically tailored output representations enable to transfer of knowledge between different platforms and allow to build generalizable

Fig. 2.1 DroNet is a lightweight sensorimotor policy that can control drones in urban environments. DroNet can follow street rules (left), fly at high altitude (center), and also avoid dynamic obstacles (right). In addition, it also generalizes to indoor environments and parking lots

sensorimotor policies. Dronet can successfully fly not only at the same height as cars, but also at relatively high altitudes, and even in indoor environments, such as parking lots and corridors.

Related Software

(S1) https://github.com/uzh-rpg/rpg_public_dronet

Related Videos

(V1) https://youtu.be/ow7aw9H4BcA

DroNet Application to Nano-Drones

The DroNet algorithm which I developed enabled the first demonstration of an autonomous nano-drone capable of closed-loop end-to-end learning-based visual navigation. Building an autonomous system at a nano-scale is very challenging since nano-drones can carry very little weight, and therefore computation: a full-stack navigation system made of state-estimation, planning, and control is out of reach of the computational budget available at these scales. Therefore, to achieve closed-loop visual navigation, we developed a complete methodology for parallel execution of DroNet directly on-board of a resource-constrained milliwatt-scale processor. Our system is based on GAP8, a novel parallel ultra-low-power computing platform, and a 27 g commercial, open-source CrazyFlie 2.0 nano-quadrotor.

(A1) Daniele Palossi, Antonio Loquercio, Francesco Conti, Eric Flamand, Davide Scaramuzza, and Luca Benini. "A 64mW DNN-based Visual Navigation Engine for Autonomous Nano-Drones". In: *IEEE Internet of Things Journal* (2019)

Related Videos

(V2) https://youtu.be/57Vy5cSvnaA

Related Software

(S1) https://github.com/pulp-platform/pulp-dronet

Fig. 2.2 PULP-DroNet: a deep learning-powered visual navigation engine that enables autonomous navigation of a pocket-size quadrotor in a previously unseen environment

Fig. 2.3 To perceive the environment and navigate to the next gate, I've generated data only with a non-photorealistic simulator. Due to the abundance of such data, generated with domain randomization **a**, the trained CNN can be deployed on a physical quadrotor without any finetuning **b**

Related Demonstrations

(D1) Daniele Palossi, Antonio Loquercio, Francesco Conti, Eric Flamand, Davide Scaramuzza, and Luca Benini. "A 64mW DNN-based Visual Navigation Engine for Autonomous Nano-Drones". In: *Demo at IROS* (2018) (Figs. 2.2)

2.1.2   Deep Drone Racing: From Simulation to Reality with Domain Randomization

(P2) Antonio Loquercio, Elia Kaufmann, René Ranftl, Alexey Dosovitskiy, Vladlen Koltun, and Davide Scaramuzza. "Deep Drone Racing: From Simulation to Reality With Domain Randomization". In: *IEEE Trans. Robotics* 36.1 (2019), pp. 1–14

Dynamically changing environments, unreliable state estimation, and operation under severe resource constraints are fundamental challenges that limit the deployment of small autonomous drones. I have addressed these challenges in the context of autonomous, vision-based drone racing in dynamic environments (Fig. 2.3). A racing drone must traverse a track with possibly moving gates at high speed. I have enabled this functionality by combining the performance of a state-of-the-art planning and control system with the perceptual awareness of a convolutional neural network (CNN). But how to collect enough data to train the CNN? A previous version of this work proposed to manually collect it in the real world [82], but the procedure is tedious and error-prone. Therefore, in this work, I proposed to train the perception system directly in simulation. Using domain randomization, the trained perception system directly generalizes to the real world. The abundance of simulated data makes our system more robust than its counterpart trained with real-world data to changes of illumination and gate appearance.

Related Publications

(R1) Elia Kaufmann*, Antonio Loquercio*, Rene Ranftl, Alexey Dosovitskiy, Vladlen Koltun, and Davide Scaramuzza. "Deep drone racing: Learning agile flight in dynamic environments". In: *Conference on Robot Learning (CoRL)*. 2018

Related Software

(S2) https://github.com/uzh-rpg/sim2real_drone_racing

Related Videos

(V3) https://youtu.be/vdxB89lgZhQ

2.1.3 Deep Drone Acrobatics

(P3) Elia Kaufmann*, Antonio Loquercio*, René Ranftl, Matthias Müller, Vladlen Koltun, and Davide Scaramuzza. "Deep Drone Acrobatics". In: *RSS: Robotics, Science, and Systems* (2020)

Performing acrobatic maneuvers with quadrotors is extremely challenging. Acrobatic flight requires high thrust and extreme angular accelerations that push the platform to its physical limits. In this work, I show that deep sensorimotor controllers trained entirely in simulation can be used for these extremely challenging navigation tasks. Specifically, I propose to learn a sensorimotor policy that enables an autonomous quadrotor to fly extreme acrobatic maneuvers with only onboard sensing and computation. I developed a theoretically sound framework for knowledge transfer between domains and found why appropriate abstractions of the visual input decrease the simulation-to-reality gap. Using this framework, I train a policy in simulation that can be directly deployed in the physical world without any fine-tuning

Fig. 2.4 A quadrotor performs a Barrel Roll (left), a Power Loop (middle), and a Matty Flip (right). We safely train acrobatic controllers in simulation and deploy them with no fine-tuning (*zero-shot transfer*) on physical quadrotors. The approach uses only onboard sensing and computation. No external motion tracking was used

on real data. This approach enables a physical quadrotor to fly maneuvers such as the Power Loop, the Barrel Roll, and the Matty Flip, during which it incurs accelerations of up to 3g (Fig. 2.4).

Related Videos

(V4) https://youtu.be/bYqD2qZJlxE

Related Software

(S3) https://github.com/uzh-rpg/deep_drone_acrobatics

2.1.4 Agile Autonomy: Learning High-Speed Flight in the Wild

(P4) Antonio Loquercio, Elia Kaufmann, René Ranftl, Matthias Müller, Vladlen Koltun, and Davide Scaramuzza. "Agile Autonomy: Learning High-Speed Flight in the Wild". In: *Science Robotics* 7 (58 2021), pp. 1–12

In this work, we use the same theoretical framework for simulation to real-world transfer developed in [81] to address one of the key problems in mobile robotics research: navigation in previously unknown cluttered environments with only on-board sensing and computation. Major industry players strive for such capabilities in their products, but current approaches [40, 122, 183] are very limited in the agility and speed they achieve in arbitrary unknown environments. The main challenge for autonomous agile flight in arbitrary environments is the coupling of fast and robust perception with effective planning. State-of-the-art works rely on systems that divide the navigation task into separate modules, i.e. sensing, mapping, and planning. Despite being successful at low speeds, these systems become brittle at higher speeds due to their modular nature, which leads to high latency, compounding errors across modules, and sensitivity to perception failures. These issues impose fundamental limits on the speed and agility that an autonomous drone can achieve. I've

Snow-covered branches   Foggy forest   Collapsed building

Fig. 2.5 Samples of man-made and natural environments where our sensorimotor policy has been deployed. All environments are previously unknown. The navigation policy was trained entirely in simulation and uses only onboard sensing and computation

departed from this modular paradigm by directly mapping noisy sensory observations to navigation commands with a specifically designed convolutional neural network. This holistic paradigm drastically reduces latency and is more robust to perception errors by leveraging regularities in the data. Similarly to my previous work [81], the approach was trained exclusively in a simplistic simulation environment. Controlled experiments in simulation show that the proposed method decreases the failure rate up to ten times in comparison to prior works. Without any adaptation, the learned controller was used to fly a real quadcopter in different challenging indoor and outdoor environments at very high speeds (Figs. 2.5 and 2.6).

Related Videos

(V5) https://youtu.be/uTWcC6IBsE4

2.1.5 Limitations of Transfer Learning via Abstraction

I have shown the validity of the proposed framework for transfer learning via abstraction in multiple works and applications: from drone racing to high-speed navigation in unstructured environments. However, the framework also comes with some limitations. First, the required abstraction function is by design domain agnostic but task-specific. Therefore, if we change the task, *e.g.* from drone racing to acrobatics, the abstraction will have to be adjusted accordingly. One possibility to do so in an automated fashion is using domain randomization, as done in my work on racing [105]. However, domain randomization requires strong simulation engineering and expensive trial and error in the real world to define the randomization bounds. Another possibility is to pre-define the abstraction function using domain knowledge, *e.g.* feature tracks [81] or depth maps [104]. This choice favors sample efficiency, simplifies training, and promotes generalization. However, the pre-defined abstraction could potentially be suboptimal for the downstream task.

Fig. 2.6 Deployment of the proposed approach in a set of challenging environments. To get a better sense of the speed and agility of the autonomous system, please watch the supplementary video

The second limitation of the proposed method is that despite abstractions, it is impossible to account for effects or motions never observed at training time. For example, if only training on planar data [106], the policy will likely not generalize to complex 3D motion. Indeed, when deploying the policies in the real world, we expect the policy to be robust to the effects observed at training time. If such effects cannot be eliminated via abstraction, *e.g.* in the case of aerodynamic drag, sudden drops in the power supply, or drifts in inertial measurements, the general practice consists of randomizing them at training time. Yet, the extent to which simulators should represent reality is unclear, and empirical investigations in this direction are exciting research avenues for future work. However, there will always be complex or too computationally intensive effects to simulate, *e.g.* contact forces or interactions with other (artificial or biological) agents. To address this limitation, I believe that the policy will require online adaptation to the environment and task. Doing so in the real world could be challenging due to the lack of privileged information or explicit reward signals but could be supported, for example, by self-supervised learning [63].

2.2 Uncertainty Estimation for Safe Deep Learning

As highlighted in the previous section, learning-based methods offer major opportunities for robust and low-latency control of drones in previously unknown environments. However, it is well known that neural network predictions are unreliable when the input sample is out of the training distribution or corrupted by noise [84]. Being able to detect such failures automatically is fundamental to integrate deep learning algorithms into robotics. Indeed, estimating uncertainties enables the interpretability of neural networks' predictions and provides valuable information during decision making [162]. This problem is even more important in the context of simulation to reality transfer, where it is not known a priori whether the training data collected in simulation is sufficient for generalization in the physical world.

Prediction uncertainty in deep neural networks generally derives from two sources: *data* uncertainty and *model* uncertainty. The former arises because of data noise, which is generally due to sensors' imperfections. The latter instead is generated from unbalances in the training data distribution. For example, a rare sample should have higher model uncertainty than a sample that appears more often in the training data. Both components of uncertainty play an important role in robotic applications. A sensor can indeed never be assumed to be noise-free, and training datasets cannot be expected to cover all the possible edge-cases.

Current approaches for uncertainty estimation require changes to the network and optimization process, typically ignore prior knowledge about the data, and tend to make over-simplifying assumptions that underestimate uncertainty. These problems have hindered their application to robotics, slowing down the integration of neural networks to large-scale, robotic systems. To address these challenges, in [103] I have proposed a novel general framework for uncertainty estimation in deep learning (Fig. 2.7).

Fig. 2.7 A neural network trained for steering angle prediction can be fully functional on a clean image (left) but generate unreliable predictions when processing a corrupted input (right). In this chapter, I propose a general framework to associate each network prediction with an uncertainty (illustrated above in red) that allows the detection of such failure cases automatically

2.2.1 A General Framework for Uncertainty Estimation in Deep Learning

(P5) Antonio Loquercio, Mattia Segù, and Davide Scaramuzza. "A General Framework for Uncertainty Estimation in Deep Learning". In: *IEEE Robotics Autom. Lett.* 5.2 (2020), pp. 3153–3160. https://doi.org/10.1109/LRA.2020. 2974682

In this paper, I present a general framework for uncertainty estimation which combines Bayesian belief networks [14, 44, 52] with Monte-Carlo sampling. Specifically, I proposed two key innovations with respect to previous works on uncertainty estimation: the use of prior information about the data, *e.g.*, sensor noise, to compute data uncertainty, and the modeling of the relationship between data and model uncertainty. These innovations enable my framework to capture prediction uncertainties better than state-of-the-art methodologies.

Due to the large number of (possibly non-linear) operations required to generate predictions, the posterior distribution $p(\mathbf{y}|\mathbf{x})$, where $\mathbf{y}$ are output predictions and $\mathbf{x}$ is the input of a neural network, is intractable. Formally, I define the total prediction uncertainty as $\sigma_{tot} = \mathrm{Var}_{p(\mathbf{y}|\mathbf{x})}(\mathbf{y})$. This uncertainty comes from two sources: data and model uncertainty. To estimate σ_{tot}, I derive a tractable approximation of $p(\mathbf{y}|\mathbf{x})$, which is general, i.e. agnostic to the architecture and learning process, and computationally feasible. My framework, summarized in Fig. 2.8, can be easily applied to all networks and already trained ones, given its invariance to the architecture or the training procedure. While not being specifically tailored to robotics, my approach's fundamental properties make it an appealing solution to learning-based perception and control algorithms, enabling them to be better integrated into robotic systems [162].

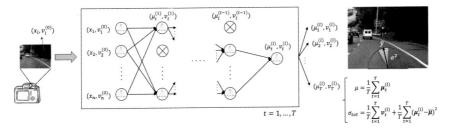

Fig. 2.8 Given an input sample **x**, associated with noise $\mathbf{v}^{(0)}$, and a trained neural network, my proposed framework computes the confidence associated with the network output. To do so, it first transforms the given network into a Bayesian belief network. Then, it uses an ensemble of T such networks, created by enabling dropout at test time, to generate the final prediction $\boldsymbol{\mu}$ and uncertainty σ_{tot}

Fig. 2.9 Application of the proposed framework for uncertainty estimation on the task of steering angle prediction (left) and objects' future motion (right). For poorly illuminated frames or for motions which can't be easily predicted (*e.g.* a dancer's legs), the network is highly uncertain about its predictions

To show the generality of our framework, we perform experiments on four challenging tasks: end-to-end steering angle prediction, obstacle future motion prediction, object recognition, and closed-loop control of a quadrotor. In these tasks, we outperform existing methodologies for uncertainty estimation by up to 23% in terms of prediction accuracy (Fig. 2.9).

Related Videos

(V6) https://youtu.be/X7n-bRS5vSM
(V7) https://youtu.be/HZOxLgsk8E0

Related Software

(S4) https://github.com/uzh-rpg/deep_uncertainty_estimation

2.2.2   Limitations of Proposed Framework for Uncertainty Estimation

The proposed approach for uncertainty estimation outperforms existing methods on several metrics and tasks. However, it also comes with some limitations. First, to extract uncertainty, a neural network needs to be converted into a bayesian one with assumed density filtering. Such conversion doubles the memory requirements (a mean and a standard deviation have to be stored for each node) and the computations required for a forward pass. In addition, the conversion can be cumbersome for complex models, particularly for recurrent ones. For example, specific activation functions or temporal aggregations might require approximations of the backward and forward passes since no closed-form solutions are known.

The second limitation of the proposed framework is that uncertainty computation requires sampling. This limitation can hinder applying the framework to the complex neural network in real-time estimation or control problems. However, I have shown that for low-dimensional problems, *e.g.* full state control of a quadrotor, a relatively small amount of samples (20) are sufficient for accurate prediction and enable real-time estimation. Recent work has shown the possibility to compute model uncertainty in a sample-free fashion [130]. However, they yet require substantial simplifications of the optimization procedure, *e.g.* linearization of the parameters' jacobian, which could lead to underestimation of the total uncertainty. Therefore, I argue that combining such methods with our approach could lead to accurate and efficient uncertainty estimation methods.

2.3   Unsupervised and Weakly-Supervised Learning of Robot Vision

Robust and effective perception is one of the keys to the integration of robots in the real world. However, state-of-the-art perception systems generally build on a supervised learning paradigm, where an expert, *e.g.* a human or an automated algorithm with privileged information, provides labels for each observation. Motivated by the astonishing capabilities of natural intelligent agents and inspired by theories from psychology, this chapter explores the idea that perception could naturally emerge from interaction with the environment. Specifically, I examine this problem on a low-level task (*i.e.* depth estimation) and a high-level task (*i.e.* moving object detection).

In [102], I introduce a novel approach to learn 3D perception from the data observed by a robot exploring an environment: images and extremely sparse depth measurements, down to even a single pixel per image. Then, in [180], I propose an adversarial contextual model for detecting moving objects in images from just video data. In both papers, only raw data is used, without any supervision coming from annotated training data or some form of hard-coded geometrical constraint. These works show that—to some extent—it is possible to learn perception tasks from the data naturally observed by a robot when interacting with its surroundings, reaching the performance, or even outperforming, the performance of traditional supervised learning.

2.3.1 Learning Depth with Very Sparse Supervision

(P6) Antonio Loquercio, A. Dosovitskiy, and D. Scaramuzza. "Learning Depth With Very Sparse Supervision". In: *IEEE Robotics and Automation Letters* 5 (2020), pp. 5542–5549

This work proposes to learn a dense depth estimation system from the data available to robots interacting with the environment: a sequence of frames and sparse depth measurements (up to one pixel per image). To learn a depth estimator from such assumptions, we design a specialized global-local deep architecture consisting of two modules, global and local. The global module takes as input two images and optical flow between them and outputs a compact latent vector of "global parameters". We expect those parameters to encode information about the observer's motion, the camera's intrinsics, and scene's features, *e.g.*, planarity. The local module then generates a compact fully convolutional network, conditioned on the "global parameters", and applies it to the optical flow field to generate the final depth estimate. The global and the local modules are trained jointly end-to-end with the available sparse depth labels and without the camera's pose or intrinsics ground truth. The proposed approach outperforms other architectural baselines and unsupervised depth estimation methods in several controlled experiments (Fig. 2.10).

Related Software

(S5) https://tinyurl.com/5kzn4r72

Related Videos

(V8) https://youtu.be/a-d40Md6tKo

2.3.2 Unsupervised Moving Object Detection via Contextual Information Separation

(P7) Yanchao Yang*, Antonio Loquercio*, Davide Scaramuzza, and Stefano Soatto. "Unsupervised Moving Object Detection via Contextual Information Separation". In: *IEEE Conference on Computer Vision and Pattern Recognition, CVPR*. 2019, pp. 879–888. https://doi.org/10.1109/CVPR.2019.00097

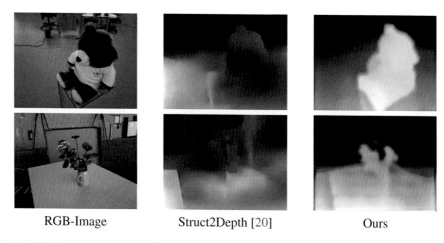

RGB-Image   Struct2Depth [20]   Ours

Fig. 2.10 I propose a novel architecture and training procedure to predict monocular depth with what would be available to a robot interacting with the environment: images and very sparse depth measurements

In this work, I propose a framework to automatically detect moving objects in a video. Being able to rapidly detect independently moving objects in a wide variety of scenes from images is functional to the survival of animals and autonomous vehicles alike. To detect moving objects, I identify parts of the image whose motion cannot be explained by that of their surroundings. In other words, the motion of the background is uninformative of the motion of the foreground and vice-versa. Although this method requires *no supervision* whatsoever, it outperforms several methods that are pre-trained on large annotated datasets. The proposed approach can be thought of as a generalization of classical variational generative region-based segmentation, but in a way that avoids explicit regularization or solution of partial differential equations at run-time (Fig. 2.11).

Related Videos

(V9) https://youtu.be/01vClieIQBw

Related Software

(S6) http://rpg.ifi.uzh.ch/unsupervised_detection.html

2.3.3 Limitations of Unsupervised or Weakly-Supervised Learning

Learning via supervision, either explicit in terms of labels or implicit in reward functions, has governed both the robotics and computer vision research communities.

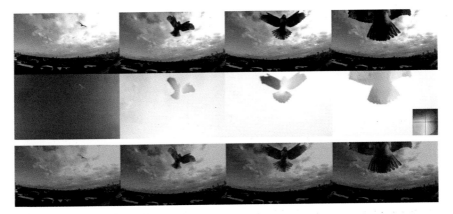

Fig. 2.11 An encounter between a hawk and a drone (top). The latter will not survive without being aware of the attack. Detecting moving objects is crucial to the survival of animal and artificial systems alike. Note that the optical flow (middle row) is quite diverse within the region where the hawk projects: It changes both in space and time. Grouping this into a moving object (bottom row) is the goal of this work

Recently such paradigm has been challenged by the concept of self-supervised and unsupervised learning, which promises to scale up the training process (no supervision needed) and improve generalization. However, especially in robotics, yet a significant gap exists between unsupervised and supervised methods.

The methods presented in this thesis make a step forward towards self-supervised learning for robotic perception. These methods are general and flexible, but they come with two main limitations. The first one is the gap of performance with supervised methods. At the time of publication, the proposed methods achieve inferior performance than supervised ones if enough supervised labeled data exists. It is reasonable to expect that more advanced learning and optimization algorithms will likely invert this trend soon. A second limitation of the proposed methods is that they are mainly effective in perception tasks, but they have not yet proven successful in training sensorimotor controllers. The biggest challenge is defining the learning signal: why should an action be considered better than another one if no supervision is present? One possible solution to this problem is back-propagation-through-time (BBTT). Such an approach propagates forward in time a (potentially learned) model of the robot in function of actions and compares the resulting predictions to the desired states. Being the model differentiable, one can optimize actions by any optimization technique, *e.g.* gradient descent. I argue that such an unsupervised learning method will achieve results that are comparable or even superior to the ones obtained by current supervised methods in the field of sensorimotor control [81, 105], while keeping the generality and flexibility of optimization-based methods [12].

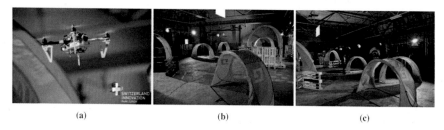

(a)   (b)   (c)

Fig. 2.12 Autonomous Drone racing demonstrations given during the inauguration of the Switzerland Innovation Park Zurich

2.4   Additional Contributions

I now present additional contributions of this thesis, which consist of a public drone racing demonstrator, as well as a set of publications unrelated to the main topic of this thesis.

2.4.1   *Drone Racing Demonstrator at Switzerland Innovation Park*

I believe that live demonstrations are one of the best ways to communicate research results while pushing to develop systems that are lightweight and work in the real world. My lab participated in the inauguration of the Switzerland Innovation Park Zurich[1] in March 2018. Drone racing with only on-board sensing and computation, one of the core contributions of this thesis, was demonstrated on that occasion in front of a general public and members of the Swiss government. In this context, we built a race track of 10 gates at different heights with length of approximately 60 m (Fig. 2.12b). The demonstration consisted of flying this track repeatedly while constantly moving gates during execution. Figure 2.12 shows some pictures captured during this event.

2.4.2   *Unrelated Contributions*

During my Ph.D., I developed a strong interest in a biologically-inspired neuromorphic vision sensor called event camera. Event cameras record asynchronous streams of per-pixel brightness changes, referred to as "events". They have appealing advantages over frame-based cameras for computer vision and robotics, including high temporal resolution, high dynamic range, and no motion blur. I believe that these properties will make them play a fundamental role in the future of mobile robotics.

---

[1] https://www.switzerland-innovation.com/zurich/.

Instead of capturing brightness images at a fixed rate, event cameras measure brightness changes (called events) for each pixel independently, preventing the use of classic computer vision algorithms. I have contributed to several works to make state-of-the-art deep learning systems accessible to event cameras. I have also applied some of these ideas in the context of 3D data processing.

Event-based Steering Angle Prediction for Self-driving Cars

(U1) Ana I. Maqueda, Antonio Loquercio, Guillermo Gallego, Narciso Garcia, and Davide Scaramuzza. "Event-Based Vision Meets Deep Learning on Steering Prediction for Self-Driving Cars". In: *IEEE Conference on Computer Vision and Pattern Recognition (CVPR)*. 2018, pp. 5419–5427. https://doi.org/10.1109/CVPR.2018.00568

This paper presents a deep neural network approach that unlocks the potential of event cameras on a challenging motion-estimation task: prediction of a vehicle's steering angle. To make the best out of this sensor-algorithm combination, we adapt state-of-the-art convolutional architectures to the output of event sensors (Fig. 2.13) and extensively evaluate the performance of our approach on a publicly available large scale event-camera dataset (≈ 1000 km). We present qualitative and quantitative explanations of why event cameras allow robust steering prediction even in cases where traditional cameras fail, *e.g.* challenging illumination conditions and fast motion.

Related Videos

(V10) https://youtu.be/_r_bsjkJTHA

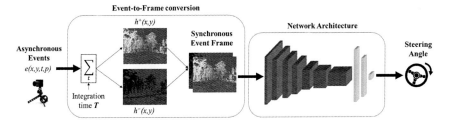

Fig. 2.13 Block diagram of the proposed approach. The output of the event camera is collected into frames over a specified time interval T, using a separate channel depending on the event polarity (positive and negative). The resulting synchronous event frames are processed by a ResNet-inspired network, which produces a prediction of the steering angle of the vehicle

Learning of Representations for Asynchronous Event-Based Data

(U2) Daniel Gehrig, Antonio Loquercio, Konstantinos G. Derpanis, and Davide Scaramuzza. "Endto- End Learning of Representations for Asynchronous Event-Based Data". In: *International Conference on Computer Vision, ICCV.* 2019, pp. 5632–5642. https://doi.org/10.1109/ICCV.2019.00573

We introduce a general framework to convert asynchronous event streams into grid-based representations through a sequence of differentiable operations (Fig. 2.14). Our framework comes with two main advantages: (i) allows learning the input event representation together with the task dedicated network in an end-to-end manner, and (ii) lays out a taxonomy that unifies the majority of extant event representations in the literature and identifies novel ones. Empirically, we show that our approach to learning the event representation end-to-end yields an improvement of approximately 12% on optical flow estimation and object recognition over state-of-the-art methods.

Related Publications

(R2) Nico Messikommer, Daniel Gehrig, Antonio Loquercio, and Davide Scaramuzza. "Event-Based Asynchronous Sparse Convolutional Networks". In: *European Conference Computer Vision (ECCV).* vol. 12353. 2020, pp. 415–431. https://doi.org/10.1007/978-3-030-58598-3_25

Related Videos

(V11) https://youtu.be/bQtSx59GXRY

Related Software

(S7) https://github.com/uzh-rpg/rpg_event_representation_learning

3D Mesh Processing

(U3) Francesco Milano, Antonio Loquercio, Antoni Rosinol, Davide Scaramuzza, and Luca Carlone. "Primal-Dual Mesh Convolutional Neural Networks". In: *Advances in Neural Information Processing Systems (NeurIPS).* 2020

We propose a novel method that allows performing inference tasks on three-dimensional geometric data by defining convolution and pooling operations on triangle meshes. We extend a primal-dual framework drawn from the graph-neural-network literature to triangle meshes and define convolutions on two types of graphs constructed from an input mesh. Our method takes features for both edges and faces of a 3D mesh as input and dynamically aggregates them using an attention mechanism. At the same time, we introduce a pooling operation with a precise geometric interpretation, that allows handling variations in the mesh connectivity by clustering

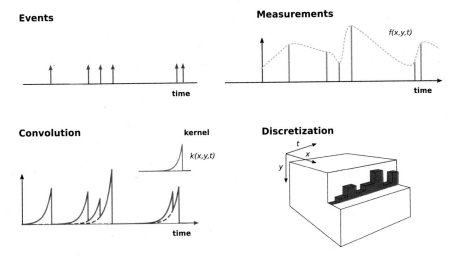

Fig. 2.14 An overview of our proposed framework. Each event is associated with a measurement (green) which is convolved with a (possibly learnt) kernel. This convolved signal is then sampled on a regular grid. Finally, various representations can be instantiated by performing projections over the temporal axis or polarities

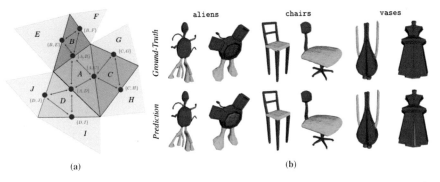

Fig. 2.15 Primal-dual graphs associated to an input mesh in PD-MeshNet **a** and qualitative results for the segmentation task on the COSEG dataset [175] **b**

mesh faces in a task-driven fashion. We provide extensive empirical results and theoretical insights of our approach using tools from the mesh-simplification literature (Fig. 2.15).

Related Videos

(V12) https://youtu.be/O8lgQgqQeuo

Related Software

(S8) https://github.com/MIT-SPARK/PD-MeshNet

Chapter 3
Future Directions

Deep learning is an emerging technology with the promise of addressing the synergy between robotic perception and action. Research on learning-based robotics is still in the early stages and has not yet reached the same maturity level as traditional model-based algorithms. Yet, considerable progress has been made in the last few years, which has clearly shown the potential of deep learning to overcome some of the limitations of conventional model-based autonomy.

A tight perception-action loop has numerous advantages (low latency, robustness to imperfect perception, quick adaptation to new conditions and tasks). However, it also comes with interesting challenges (sample complexity, interpretability, use of domain knowledge, generalization) which this aimed to address. Using the demonstrator of high-speed agile flight, I have shown that a tight perception-action loop enables operation in new scenarios previously inaccessible to quadrotors. I now outline a few promising research directions to address the limitations of sample complexity and generalizability of the proposed methods.

Beyond Quadrotor Flight. The main demonstrator of the proposed methodologies has been agile quadrotor flight. However, the contributions of this thesis go beyond this application. Indeed, the idea of learning through input and output abstraction is a general methodology for knowledge transfer. Such an idea can be particularly effective when the input and output space is high dimensional: not only can abstractions narrow the gap between domains but also they can simplify training and favor generalization. Indeed, the idea of abstraction has been recently proved successful on both ground robots and manipulators [146, 185]. The main limitation of the approach is the lack of generalization to effects that cannot be modeled or abstracted away. For example, one challenge is to build policies for agent-to-agent or agent-to-human interaction. Simulation to reality transfer has therefore experienced only limited success in this domain. One possible solution that I envision is sample efficient adaption to the environment and task (more on this is discussed below).

Similarly, the proposed uncertainty estimation framework is not limited to the computer vision and control tasks presented in this thesis. Indeed, I argue that all

© The Author(s), under exclusive license to Springer Nature Switzerland AG 2023   43
A. Loquercio, *Agile Autonomy: Learning High-Speed Vision-Based Flight*, Springer Tracts in Advanced Robotics 153, https://doi.org/10.1007/978-3-031-27288-2_3

safety-critical applications of machine learning can benefit from it. Due to the complexity and variability of the real world, eliminating the errors in neural network predictions due to noise in the inputs or out-of-distribution testing is impossible. Such errors should therefore be detected and used for continual policy improvement. Yet, the main limitation lies in its computational efficiency, which can be prohibitively high for complex models with high-dimensional input and/or output spaces. Such limitation can be addressed by combining my framework with recent methods for sampling-free uncertainty estimation [130, 151].

Online Adaptive Learning. Natural agents operate in a radically different way than artificial agents. For the former, the concept of *training data* does not exist: they continuously learn to cope with their senses and experiences in the real world. In contrast, the behaviors of artificial agents are generally defined in advance and remain largely unchanged during operation. Building into artificial agents the ability to adapt could enable the efficient generalization of their knowledge to new tasks and conditions. Recent research has developed new machine learning algorithms to enable few-shot learning to new tasks and conditions, *e.g.* meta learning [5, 39, 132] or adaptive learning [86, 107, 148]. While their application in controlled environments and standard visual tasks shows very promising results, it is still unclear to what extent they can be applied to more challenging, real-world robotic tasks.

Causal Inference. The process of determining the independent effects of a particular phenomenon (cause) in a system is generally referred to as causal inference. Humans use causal cues and their related effects to make both short-term and long-term predictions, to understand the mechanism that leads to changes in the system, and to support decision making [62]. Can causal reasoning be similarly beneficial for robots? Arguably, it would enable the development of more complex representations of the world, favor adaptation and generalization to previously unseen scenarios. In addition, it could naturally provide a transparent decision-making process. For instance, a robot could understand that the effects of crashing into objects are always negative, independently of the context and application, without explicitly forcing this behavior in the training process(*e.g.* with domain randomization). Causal inference has had a large impact in several disciplines, including computer science [127], economics [172], and epidemiology [139]. Very recently causal inference also made its first steps into computer vision [145] and robotics [91], leading to increased robustness in the former and computational efficiency in the latter. However, it is not yet clear how to build causal structures from unstructured data, and how to efficiently use causal cues to support decision making in complex robotics tasks. I believe that the interactive nature of perception and action can help to solve this problem—for example, it could provide structure to the sensor data observed during operation.

Social Learning. The process of acquiring skills through the interaction with other agents is generally defined as *social learning*. Some of the most unique human cognitive skills, *e.g.* linguistic communication, abstract reasoning, and complex motor skills, are rooted in social learning [59]. Given its importance for humans, social learning has also started to play a fundamental role in machine learning and robotics. *Self-play* is arguably its most popular incarnation: in a reinforcement learning setting, agents improve by competing against each other [7, 154]. However, social learning is not all about competition: very recent research has shown its potential for achieving cooperation and coordination between artificial agents [75], developing efficient inter-communication schemes [176], and improving open-domain conversational agents [54]. Given the technical difficulties to build and control multiple robots in the real world, social learning is still in the early stages of robotics. Most of the current research has focused on the problem of human-robot interaction [58] or information exchange minimization [70]. Nonetheless, I believe that social learning could also benefit from the interactive nature of robots, powered by a tight perception-action loop, and lead to numerous advantages, *e.g.* improving the learning speed of new tasks, acquiring more robust motor skills, or exploiting unstructured social guidance typically offered by humans.

Beyond traditional visual sensors. Robots strongly rely on vision-based sensors, *e.g.* RGB cameras, to understand and interact with their surroundings. While other alternatives exist, *e.g.* distance-based sensors like lidar or radar, they have a lower resolution than cameras, are generally bulky and more expensive, and are strongly affected by environmental factors like reflective surfaces or bad weather. In contrast, cameras have very high spatial resolution and are generally agnostic to environmental features, which made them popular in robotics. However, despite their advantages, traditional cameras also have several non-idealities. For example, they are subject to motion-blur, low dynamic range, and low temporal resolution, which makes them unreliable when navigating at high-speed or in high-dynamic-range scenarios (*e.g.* when moving from indoor to outdoor). Novel bio-inspired vision sensors could be complementary to traditional cameras, offering the ability to still sense the environment in those extreme conditions. Event cameras are one of them: in contrast to frame-based cameras, they record asynchronous streams of per-pixel brightness changes, referred to as "events". They have appealing advantages over frame-based cameras for computer vision and robotics, including high temporal resolution, high dynamic range, and no motion blur. The research on event cameras is still in its early stages compared to traditional computer vision [49]. Still, already several robotic applications have shown benefits of event cameras, like flying in the dark [137], avoiding dynamic obstacles [36], and flying despite robot failure [161]. Alternative bio-inspired imaging sensors have also recently emerged, such as cellular processor arrays (SCAMP [19]), in which every pixel has a processor that allows performing several types of computations with the brightness of the pixel and its neighbors. These novel sensors, combined with specifically tailored learning algorithms (*e.g.* Spiking Neural Networks [149]) and neuromorphic processing units (*e.g.* the Intel Loihi [27]), have the potential to significantly decrease the perception latency. Lower perception latency can result in a faster perception-action loop, which in turn can enable to perform complex tasks at high speeds in challenging scenarios.

References

1. P. Abbeel, A. Coates, A.Y. Ng, Autonomous helicopter aerobatics through apprenticeship learning. Int. J. Robot. Res. **29**(13), 1608–1639 (2010)
2. A. Achille, S. Soatto, *Where is the Information in a Deep Neural Network?* ArXiv preprint arXiv:1905.12213 (2019)
3. P. Agrawal et al., Learning to poke by poking: Experiential learning of intuitive physics, in *International Conference on Neural Information Processing Systems (NIPS)* (2016), pp. 5092–5100
4. R. Allen, M. Pavone, A real-time framework for kinodynamic planning with application to quadrotor obstacle avoidance, in *AIAA Guidance, Navigation, and Control Conference* (2016), p. 1374
5. M. Andrychowicz et al., Learning to learn by gradient descent by gradient descent, in *Advances in Neural Information Processing Systems (NIPS)* (2016), pp. 3981–3989
6. A. Antonini et al., The blackbird UAV dataset. Int. J. Robot. Res. **39**(10–11), 1346–136 (2020)
7. B. Baker et al., Emergent tool use from multi-agent autocurricula, in *International Conference on Learning Representations* (2020)
8. A.J. Barry, P.R. Florence, R. Tedrake, High-speed autonomous obstacle avoidance with push-broom stereo. J. Field Robot. **35**(1), 52–68 (2018). https://doi.org/10.1002/rob.21741
9. D. Beymer et al., A real-time computer vision system for measuring traffic parameters. IEEE Conference on Computer Vision and Pattern Recognition (CVPR) (1997). https://doi.org/10.1109/cvpr.1997.609371
10. M. Blösch et al., Vision based MAV navigation in unknown and unstructured environments. IEEE Int. Conf. Robot. Autom. (ICRA) 21–28 (2010)
11. C. Blundell et al., Weight uncertainty in neural networks, in *International Conference on Machine Learning* (2015), pp. 1613–1622
12. F. Borrelli, A. Bemporad, M. Morari, *Predictive Control for Linear and Hybrid Systems.* (Cambridge University Press, 2017)
13. K. Bousmalis et al., Using simulation and domain adaptation to improve efficiency of deep robotic grasping. IEEE Int. Conf. Robot. Autom. (ICRA) (2018)
14. K. Boyen, Tractable inference for complex stochastic processes, in *InUAI* (1998), pp. 33–42
15. G.J. Brostow, R. Cipolla, Unsupervised bayesian detection of independent motion in crowds, in *IEEE Conference on Computer Vision and Pattern Recognition (CVPR)* (IEEE, 2006). https://doi.org/10.1109/cvpr.2006.320.
16. T. Brox, A. Bruhn, J. Weickert, Variational motion segmentation with level sets, in *IEEE European Conference on Computer Vision (ECCV)* (2006), pp. 471–483. https://doi.org/10.1007/11744023_37.

© The Editor(s) (if applicable) and The Author(s), under exclusive license to
Springer Nature Switzerland AG 2023
A. Loquercio, *Agile Autonomy: Learning High-Speed Vision-Based Flight*, Springer
Tracts in Advanced Robotics 153, https://doi.org/10.1007/978-3-031-27288-2

17. A. Bry, A. Bachrach, N. Roy, State estimation for aggressive flight in GPS-denied environments using onboard sensing, in *International Conference on Robotics and Automation* (IEEE, 2012), pp. 1–8
18. A. Bry, N. Roy, Rapidly-exploring Random Belief Trees for motion planning under uncertainty. IEEE Int. Conf. Robot. Autom. (ICRA) (2011)
19. S.J. Carey et al., A 100,000 fps vision sensor with embedded 535 GOPS/W 256x256 SIMD processor array, in *VLSI Circuits Symp* (2013)
20. V. Casser et al., Depth prediction without the sensors: Leveraging structure for unsupervised learning from monocular videos, in *Thirty-Third AAAI Conference on Artificial Intelligence (AAAI-19)* (2019), pp. 8001–8008
21. B. Cheng et al., Panoptic-DeepLab: A simple, strong, and fast baseline for bottom-up panoptic segmentation, in *CVPR* (2020)
22. J. Cheng et al., SegFlow: Joint learning for video object segmentation and optical flow, in *IEEE International Conference on Computer Vision (ICCV)* (2017)
23. K. Chua et al., Deep reinforcement learning in a handful of trials using probabilistic dynamics models, in *Advances in Neural Information Processing Systems* (2018), pp. 4754–4765
24. T. Cieslewski, E. Kaufmann, D. Scaramuzza, Rapid exploration with multi-rotors: A frontier selection method for high speed flight. IEEE/RSJ Int. Conf. Intell. Robot. Syst. (IROS) 2135–2142 (2017)
25. M. Cordts et al., The cityscapes dataset for semantic urban scene understanding, in *IEEE Conference on Computer Vision and Pattern Recognition (CVPR)* (2016)
26. D. Cremers, S. Soatto, Motion competition: A variational approach to piecewise parametric motion segmentation. IEEE International Journal of Computer Vision **62**(3), 249–265 (2004). https://doi.org/10.1007/s11263-005-4882-4
27. M. Davies et al., Loihi: A neuromorphic manycore processor with on-chip learning. IEEE Micro **38**(1), 82–99 (2018)
28. J. Deng et al., ImageNet: A large-scale image database, in *2009 IEEE Conference on Computer Vision and Pattern Recognition* (2009), pp. 248–255
29. A. Dosovitskiy et al., CARLA: An open urban driving simulator, in *Conference on Robot Learning (CORL)* (2017), pp. 1–16
30. D. Eigen, C. Puhrsch, R. Fergus, Depth map prediction from a single image using a multi-scale deep network. Conf. Neural Inf. Process. Syst. (NIPS) 2366–2374 (2014)
31. A. Elnakib et al., Medical image segmentation: a brief survey, in *Multi Modality State-of-the-Art Medical Image Segmentation and Registration Methodologies* (Springer, 2011), pp. 1–39
32. J. Engel, V. Koltun, D. Cremers, Direct sparse odometry. IEEE Trans. Pattern Anal. Mach. Intell. (T-PAMI) **40**(3), 611–625 (2018)
33. S.M. Ali Eslami et al., Neural scene representation and rendering. Science **360**(6394), 1204–1210 (2018)
34. M. Faessler et al., Autonomous, vision-based flight and live dense 3D mapping with a quadrotor micro aerial vehicle. J. Field Robot. **33**(4), 431–450 (2016)
35. D. Falanga, S. Kim, D. Scaramuzza, How fast is too fast? the role of perception latency in high-speed sense and avoid. IEEE Robot. Autom. Lett. **4**(2), 1884–1891 (Apr. 2019). ISSN: 2377–3766. https://doi.org/10.1109/LRA.2019.2898117.
36. D. Falanga, K. Kleber, D. Scaramuzza, Dynamic obstacle avoidance for quadrotors with event cameras. Sci. Robot. **5**(40) (2020)
37. D. Falanga et al., Aggressive quadrotor flight through narrow gaps with onboard sensing and computing using active vision. IEEE Int. Conf. Robot. Autom. (ICRA) (2017)
38. D. Feng, L. Rosenbaum, K. Dietmayer, Towards safe autonomous driving: Capture uncertainty in the deep neural network for lidar 3D vehicle detection, in *Proceedings of the 21st IEEE International Conference on Intelligent Transportation Systems* (2018)
39. C. Finn, P. Abbeel, S. Levine, Model-agnostic meta-learning for fast adaptation of deep networks, in *International Conference on Machine Learning (ICML)*, ed. by Doina, P., Yee, W.T. vol. 70. Proceedings of Machine Learning Research (2017), pp. 1126–1135

40. P. Florence, J. Carter, R. Tedrake, Integrated perception and control at high speed: Evaluating collision avoidance maneuvers without maps, in *Algorithmic Foundations of Robotics XII* (Springer, 2020), pp. 304–319

41. P. Foehn et al., AlphaPilot: Autonomous drone racing, in *Proceedings of Robotics: Science and Systems* (Corvalis, Oregon, USA, July 2020). https://doi.org/10.15607/RSS.2020.XVI. 081.

42. C. Forster et al., SVO: Semidirect visual odometry for monocular and multicamera systems. IEEE Trans. Robot. **33**(2), 249–265 (2017). https://doi.org/10.1109/TRO.2016.2623335

43. M. Fraccaro et al., A disentangled recognition and nonlinear dynamics model for unsupervised learning, in *Advances in Neural Information Processing Systems* (2017), pp. 3601–3610

44. F. Hinton, Variational learning in nonlinear Gaussian belief networks. Neural Comput. **11**(1), 193–213 (1999)

45. F. Furrer et al., RotorS-A modular gazebo MAV simulator framework, in *Robot Operating System (ROS)* (Springer, 2016), pp. 595–625

46. S. Saeedi G. et al., 3D mapping for autonomous quadrotor aircraft. Unmanned Syst. **5**(3), 181–196 (2017). https://doi.org/10.1142/S2301385017400064

47. Y. Gal, Z. Ghahramani, Dropout as a bayesian approximation: Representing model uncertainty in deep learning, in *Proceedings of the 33rd International Conference on Machine Learning* (2015)

48. Y. Gal, J. Hron, A. Kendall, Concrete dropout, in *Advances in Neural Information Processing Systems* (2017), pp. 3581–3590

49. G. Gallego et al., *Event-based vision: A survey* (IEEE Trans. Pattern Anal. Mach, Intell, 2020)

50. D. Gandhi, L. Pinto, A. Gupta, Learning to fly by crashing, in *International Conference on Intelligent Robots and Systems, IROS* (2017), pp. 3948–3955

51. R. Garg et al., Unsupervised CNN for single view depth estimation: Geometry to the rescue. Eur. Conf. Comput. Vis. (ECCV). 740–756 (2016)

52. J. Gast, S. Roth, Lightweight probabilistic deep networks, in *Proceedings of the IEEE Conference on Computer Vision and Pattern Recognition* (2018)

53. D. Gehrig et al., End-to-end learning of representations for asynchronous event- based data, in *International Conference on Computer Vision, ICCV* (2019), pp. 5632–5642. https://doi.org/10.1109/ICCV.2019.00573.

54. A. Ghandeharioun et al., *Approximating Interactive Human Evaluation with Self-Play for Open-Domain Dialog Systems* (2019), pp. 13658–13669

55. J. James, *Gibson* (The Ecological Approach to Visual Perception, Houghton Mifflin, 1979)

56. C. Godard, O.M. Aodha, G.J. Brostow, Unsupervised monocular depth estimation with left-right consistency. IEEE Int. Conf. Comput. Vis. Pattern Recog. (CVPR). 6602–6611 (2017)

57. A. Gordon et al., Depth from videos in the wild: Unsupervised monocular depth learning from unknown cameras. Int. Conf. Comput. Vis. (ICCV). 8977–8986 (2019)

58. J. de Greeff, T. Belpaeme, Why robots should be social: Enhancing machine learning through social human-robot interaction. PLOS ONE **10**(9) (2015)

59. J.E. Grusec, Social learning theory and developmental psychology: The legacies of Robert R. Sears and Albert Bandura, in *A Century of Developmental Psychology* (American Psychological Association, 1994), pp. 473–497

60. W. Guerra et al., FlightGoggles: Photorealistic sensor simulation for perceptiondriven robotics using photogrammetry and virtual reality. IEEE/RSJ Int. Intell. Robot. Syst. (IROS) (2019)

61. A. Gupta et al., Learning invariant feature spaces to transfer skills with reinforcement learning, in *Internation Conference on Learning Representation (ICLR)* (2017)

62. Y. Hagmayer et al., Causal reasoning through intervention, in *Causal Learning.* (Oxford University Press, 2007), pp. 86–100

63. N. Hansen et al., Self-supervised policy adaptation during deployment. arXiv preprint arXiv:2007.04309 (2020)

64. R. Hartley, A. Zisserman, *Multiple View Geometry in Computer Vision*, 2nd edn. (Cambridge University Press, 2003)

65. R.I. Hartley, In defense of the eight-point algorithm. IEEE Trans. Pattern Anal. Mach. Intell. **19**(6), 580–593 (1997)

66. R.I. Hartley, P.F. Sturm, Triangulation. Comput. Vis. Image Underst. **68**(2), 146–157 (1997)

67. K. He et al., Deep residual learning for image recognition, in *Proceedings of the IEEE Conference on Computer Vision and Pattern Recognition* (2016), pp. 770–778

68. L. Heng et al., Autonomous visual mapping and exploration with a micro aerial vehicle. J. Field Robot. **31**(4), 654–675 (2014)

69. J.M. Hernández-Lobato, R. Adams, Probabilistic backpropagation for scalable learning of bayesian neural networks, in *International Conference on Machine Learning* (2015), pp. 1861–1869

70. A. Hock, A.P. Schoellig, Distributed iterative learning control for multiagent systems. Auton. Robots **43**(8), 1989–2010 (2019)

71. G. Huang et al., Densely connected convolutional networks, in *IEEE Conference on Computer Vision and Pattern Recognition (CVPR)* (2017), pp. 2261–2269. https://doi.org/10.1109/CVPR.2017.243.

72. J. Hwangbo, J. Lee, M. Hutter, Per-contact iteration method for solving contact dynamics. IEEE Robot. Autom. Lett. **3**(2), 895–902 (2018)

73. O. Ian, Risk versus uncertainty in deep learning: Bayes, bootstrap and the dangers of dropout, in *Advances in Neural Information Processing Systems Workshops* (2016)

74. S. James, A.J. Davison, E. Johns, Transferring end-to-end visuomotor control from simulation to real world for a multi-stage task, in *Conference on Robot Learning (CoRL)* (2017)

75. N. Jaques et al., Social influence as intrinsic motivation for multi-agent deep reinforcement learning, in *International Conference on Machine Learning(ICML)*, vol. 97. (2019), pp. 3040–3049

76. M. Johnson-Roberson et al., Driving in the matrix: Can virtual worlds replace human-generated annotations for real world tasks? IEEE Int. Conf. Robot. Autom. (ICRA). (2017)

77. S. Jung et al., Perception, guidance, and navigation for indoor autonomous drone racing using deep learning. IEEE Robotics Autom. Lett. **3**(3), 2539–2544 (2018). https://doi.org/10.1109/LRA.2018.2808368

78. G. Kahn et al., *Uncertainty-Aware Reinforcement Learning for Collision Avoidance*. Arxiv Preprint (2017)

79. N.R. Kapania, *Trajectory Planning and Control for an Autonomous Race Vehicle*. PhD Thesis. (Stanford University, 2016)

80. E. Kaufmann et al., Beauty and the beast: Optimal methods meet learning for drone racing. IEEE Int. Conf. Robot. Autom. (ICRA). (2019)

81. E. Kaufmann et al., Deep drone acrobatics, in *RSS: Robotics, Science, and Systems* (2020)

82. E. Kaufmann et al., Deep drone racing: Learning agile flight in dynamic environments, in *Conference on Robot Learning (CoRL)* (2018)

83. J.C. Kegelman, L.K. Harbott, J. Christian Gerdes, Insights into vehicle trajectories at the handling limits: analysing open data from race car drivers. Veh. Syst. Dyn. **55**(2), 191–207 (2017)

84. A. Kendall, Y. Gal, What uncertainties do we need in bayesian deep learning for computer vision?, in *31st Conference on Neural Information Processing Systems (NIPS 2017)* (2017)

85. L. Keuntaek et al., *Ensemble Bayesian Decision Making with Redundant Deep Perceptual Control Policies*. ArXiv (2018)

86. H. Kim, A. Ng, Stable adaptive control with online learning, in *Advances in Neural Information Processing Systems*, vol. 17, eds. by L. Saul, Y. Weiss, L. Bottou (MIT Press, 2005). https://proceedings.neurips.cc/paper/2004/file/8e68c3c7bf14ad0bcaba52babfa470bd-Paper.pdf

87. N. Koenig, A. Howard, Design and use paradigms for gazebo, an opensource multi-robot simulator, in *2004 IEEE/RSJ International Conference on Intelligent Robots and Systems (IROS)*, vol. 3. (IEEE Cat. No. 04CH37566). (IEEE, 2004), pp. 2149–2154

88. S. Kohlbrecher et al., Hector open source modules for autonomous mapping and navigation with rescue robots, in *Robot Soccer World Cup. Springer* (2013), pp. 624–631

89. K. Kritayakirana, J. Christian Gerdes, Autonomous vehicle control at the limits of handling. Int. J. Veh. Auton. Syst. **10**(4), 271–296 (2012)
90. B. Lakshminarayanan, A. Pritzel, C. Blundell, Simple and scalable predictive uncertainty estimation using deep ensembles, in *Advances in Neural Information Processing Systems* (2017), pp. 6402–6413
91. K. Leung, N. Aréchiga, M. Pavone, Back-propagation through signal temporal logic specifications: Infusing logical structure into gradient-based methods. ArXiv preprint arXiv:2008.00097 (2020)
92. S. Levine et al., End-to-end training of deep visuomotor policies. J. Mach. Learn. Res. **17**, 39:1–39:40 (2016)
93. S. Li et al., *Autonomous Drone Race: A Computationally Efficient Vision-Based Navigation and Control Strategy*. ArXiv preprint arXiv:1809.05958 (2018)
94. S. Li et al., *Aggressive Online Control of a Quadrotor via Deep Network Representations of Optimality Principles*. arXiv:1912.07067 (2019)
95. S. Li et al., Visual model-predictive localization for computationally efficient autonomous racing of a 72-g drone. J. Field Robotics **37**(4), 667–692 (2020). https://doi.org/10.1002/rob.21956
96. T.-Y. Lin et al., Microsoft COCO: Common objects in context, in *European Conference on Computer Vision (ECCV)* (2014)
97. S. Liu et al., Search-based motion planning for aggressive flight in se (3). IEEE Robotics Autom. Lett. **3**(3), 2439–2446 (2018)
98. L. Ljung, S. Gunnarsson, Adaptation and tracking in system identification-A survey. Automatica **26**(1), 7–21 (1990)
99. G. Loianno, D. Scaramuzza, Special issue on future challenges and opportunities in vision-based drone navigation. J. Field Rob. **37**(4), 495–496 (2020)
100. G. Loianno et al., Estimation, control, and planning for aggressive flight with a small quadrotor with a single camera and IMU. IEEE Rob. Autom. Lett. **2**(2), 404–411 (2016)
101. H.C. Longuet-Higgins, A computer algorithm for reconstructing a scene from two projections. Nature **293**, 133–135 (1981)
102. A. Loquercio, A. Dosovitskiy, D. Scaramuzza, Learning depth with very sparse supervision. IEEE Rob. Autom. Lett. **5**, 5542–5549 (2020)
103. A. Loquercio, M. Segù, D. Scaramuzza, A general framework for uncertainty estimation in deep learning. IEEE Robo. Autom. Lett. **5**(2), 3153–3160 (2020). https://doi.org/10.1109/LRA.2020.2974682
104. A. Loquercio et al., Agile autonomy: Learning high-speed flight in the wild. Sci. Rob. **7**(58), 1–12 (2021)
105. A. Loquercio et al., Deep drone racing: From simulation to reality with domain randomization. IEEE Trans. Rob. **36**(1), 1–14 (2019)
106. A. Loquercio et al., DroNet: Learning to fly by driving. IEEE Rob. Autom. Lett. **3**(2), 1088–1095 (2018)
107. S. Lupashin et al., A simple learning strategy for high-speed quadrocopter multiflips, in *IEEE International Conference on Robotics and Automation (ICRA)* (2010), pp. 1642–1648
108. S. Lynen et al., Get out of my lab: Large-scale, real-time visual-inertial localization, in *Robotics: Science and Systems XI* (July 2015). https://doi.org/10.15607/rss.2015.xi.037.
109. R.M. MartinWicke, A. Angelova, Unsupervised learning of depth and ego-motion from monocular video using 3D geometric constraints. IEEE Int. Conf. Comput. Vis. Pattern Recog. (CVPR). 5667–5675 (2018)
110. A.I. Maqueda et al., Event-based vision meets deep learning on steering prediction for self-driving cars. IEEE Conf. Comput. Vis. Pattern Recog. (CVPR). 5419–5427 (2018). https://doi.org/10.1109/CVPR.2018.00568.
111. N. Messikommer et al., Event-based asynchronous sparse convolutional networks. Eur. Conf. Comput. Vis. (ECCV) **12353**, 415–431 (2020). https://doi.org/10.1007/978-3-030-58598-3_25

112. F. Milano et al., Primal-dual mesh convolutional neural networks, in *Advances in Neural Information Processing Systems (NeurIPS)* (2020)
113. V. Mnih et al., Human-level control through deep reinforcement learning. Nature **518**(7540), 529–533 (2015)
114. K. Mohta et al., Fast, autonomous flight in GPS-denied and cluttered environments. J. Field Robot. **35**(1), 101–120 (2018). https://doi.org/10.1002/rob.21774
115. B. Morrell et al., Differential flatness transformations for aggressive quadrotor flight. IEEE Int. Conf. Robot. Autom. (ICRA). (2018)
116. M.W. Mueller, M. Hehn, R. D'Andrea, A computationally efficient algorithm for state-to-state quadcopter trajectory generation and feasibility verification. IEEE/RSJ Int. Conf. Intell. Robot. Syst. (IROS). (2013)
117. M. Müller et al., Driving policy transfer via modularity and abstraction, in *Conference on Robot Learning* (2018), pp. 1–15
118. M. Müller et al., Sim4cv: A photo-realistic simulator for computer vision applications. Int. J. Comput. Vis. **126**(9), 902–919 (2018)
119. R. Mur-Artal, J.M.M. Montiel, J.D. Tardós, ORB-SLAM: A versatile and accurate monocular SLAM system. IEEE Trans. Robot. **31**(5), 1147–1163 (2015)
120. L. Neumann, A. Zisserman, A. Vedaldi, Relaxed softmax: Efficient confidence auto-calibration for safe pedestrian detection, in *Advances in Neural Information Processing Systems Workshops* (2018)
121. P. Ochs, J. Malik, T. Brox, Segmentation of moving objects by long term video analysis. IEEE Trans. Pattern Anal. Mach. Intell. (PAMI) **36**(6), 1187–1200 (2014). https://doi.org/10.1109/tpami.2013.242
122. H. Oleynikova et al., Continuous-time trajectory optimization for online UAV replanning. IEEE/RSJ Int. Conf. Intell. Robot. Syst. (IROS) (2016)
123. H. Oleynikova et al., Voxblox: Incremental 3D euclidean signed distance fields for on-board MAV planning. IEEE/RSJ Int. Conf. Intell. Robot. Syst. (IROS) 1366–1373 (2017)
124. T. Ozaslan et al., Autonomous navigation and mapping for inspection of penstocks and tunnels with MAVs. IEEE Robot. Autom. Lett. **2**(3), 1740–1747 (2017)
125. D. Palossi et al., A 64mW DNN-based visual navigation engine for autonomous nano-drones, in *Demo at IROS* (2018)
126. D. Palossi et al., A 64mW DNN-based visual navigation engine for autonomous nano-drones. IEEE IoT J. (2019)
127. J. Pearl et al., Causal inference in statistics: An overview. Stat. Surv. **3**, 96–146 (2009)
128. X.B. Peng et al., Sim-to-real transfer of robotic control with dynamics randomization, in *International Conference on Robotics and Automation (ICRA)* (2018), pp. 3803–3810
129. F. Perazzi et al., A benchmark dataset and evaluation methodology for video object segmentation. IEEE Conference on Computer Vision and Pattern Recognition (CVPR) (2016). https://doi.org/10.1109/cvpr.2016.85
130. J. Postels et al., Sampling-free epistemic uncertainty estimation using approximated variance propagation, in *Proceedings of the IEEE/CVF International Conference on Computer Vision (ICCV)* (Oct. 2019)
131. R. Ranftl et al., *Towards Robust Monocular Depth Estimation: Mixing Datasets for Zero-Shot Cross-Dataset Transfer*. arXiv:1907.01341 (2019)
132. S. Ravi, H. Larochelle, Optimization as a model for few-shot learning, in *International Conference on Learning Representations (ICLR)* (2017)
133. J. Redmon, A. Farhadi, YOLO9000: Better, Faster, stronger, in *Conference on Computer Vision and Pattern Recognition (CVPR)* (2017), pp. 6517–6525. https://doi.org/10.1109/CVPR.2017.690
134. D.J. Rezende et al., Unsupervised learning of 3D structure from images. Conf. Neural Inf. Process. Syst. (NIPS). 4997–5005 (2016)
135. C. Richter, A. Bry, N. Roy, Polynomial trajectory planning for aggressive quadrotor flight in dense indoor environments. Proc. Int. Symp. Robot. Res. (ISRR) (2013)

136. S.R. Richter et al., Playing for data: Ground truth from computer games, in *European Conference on Computer Vision (ECCV)* (2016)
137. A. Rosinol Vidal et al., Ultimate sLAM? Combining events, images, and IMU for robust visual SLAM in HDR and high speed scenarios. IEEE Robot. Autom. Lett. **3**(2), 994–1001 (Apr. 2018). https://doi.org/10.1109/LRA.2018.2793357.
138. S. Ross et al., Learning monocular reactive UAV control in cluttered natural environments. IEEE Int. Conf. Robot. Autom. (ICRA) 1765–1772 (2013)
139. K.J. Rothman, S. Greenland, Causation and causal inference in epidemiology. Am. J. Pub. Health **95**(S1), S144–S150 (2005)
140. A.A. Rusu et al., Sim-to-real robot learning from pixels with progressive nets, in *Conference on Robot Learning (CoRL)* (2017)
141. M. Ryll et al., Efficient trajectory planning for high speed flight in unknown environments, in *2019 International Conference on Robotics and Automation (ICRA)* (IEEE, 2019), pp. 732–738
142. G. Williams et al., Aggressive driving with model predictive path integral control, in *2016 IEEE Int. Conf. Robot. Autom. (ICRA)* (Stockholm, Sweden, May 2016), pp. 1433–1440
143. F. Sadeghi, S. Levine, CAD2RL: Real single-image flight without a single real image, in *Robotics: Science and Systems RSS*, ed. by N.M. Amato et al. (2017)
144. F. Sadeghi et al., Sim2Real viewpoint invariant visual servoing by recurrent control. Conference on Computer Vision and Pattern Recognition (CVPR) (2018). https://doi.org/10.1109/cvpr.2018.00493
145. A. Sauer, A. Geiger, Counterfactual generative networks, in *International Conference on Learning Representations* (2021)
146. A. Sax et al., Learning to navigate using mid-level visual priors, *Conference on Robot Learning* (PMLR. 2020), pp. 791–812
147. D. Scaramuzza et al., Vision-controlled micro flying robots: From system design to autonomous navigation and mapping in GPS-denied environments. IEEE Robot. Autom. Mag. **21**(3), 26–40 (2014)
148. A.P. Schoellig, F.L. Mueller, R. D'Andrea, Optimization-based iterative learning for precise quadrocopter trajectory tracking. Auton. Robot. **33**(1–2), 103–127 (2012)
149. R. Serrano-Gotarredona et al., CAVIAR: A 45k neuron, 5M synapse, 12G connects/s AER hardware sensory-processing-learning-actuating system for high-speed visual object recognition and tracking. IEEE Trans. Neural Netw. **20**(9), 1417–1438 (2009). https://doi.org/10.1109/TNN.2009.2023653
150. S. Shah et al., Airsim: High-fidelity visual and physical simulation for autonomous vehicles, in *Field and Service Robot* (Springer, 2018), pp. 621–635
151. A. Sharma, N. Azizan, M. Pavone, *Sketching Curvature for Efficient Out-of-Distribution Detection for Deep Neural Networks* (2021). arXiv: 2102.12567
152. S. Shen et al. Vision-based state estimation and trajectory control towards high-speed flight with a quadrotor, in *Robotics: Science and Systems (RSS)* (2013)
153. J. Shi, J. Malik, Motion segmentation and tracking using normalized cuts. IEEE International Conference on Computer Vision (ICCV) (1998). https://doi.org/10.1109/iccv.1998.710861
154. D. Silver et al., Mastering the game of go without human knowledge. Nature **550**(7676), 354–359 (2017)
155. H. Song et al., Pyramid dilated deeper ConvLSTM for video salient object detection. IEEE European Conference on Computer Vision (ECCV) (2018). https://doi.org/10.1007/978-3-030-01252-6_44
156. Y. Song et al., Flightmare: A flexible quadrotor simulator, in *Conference on Robot Learning* (2020)
157. R. Spica et al., A real-time game theoretic planner for autonomous two-player drone racing. IEEE Trans. Robot. **36**(5), 1389–1403 (2020)
158. F. Stulp et al., Learning to grasp under uncertainty, in *2011 IEEE International Conference on Robotics and Automation* (IEEE, 2011), pp. 5703–5708

159. Q. Su et al., Nonlinear statistical learning with truncated gaussian graphical models, in *International Conference on Machine Learning* (2016), pp. 1948–1957

160. D. Sun et al., A fully-connected layered model of foreground and background flow. IEEE Conference on Computer Vision and Pattern Recognition (CVPR) (2013). https://doi.org/10.1109/cvpr.2013.317

161. S. Sun et al., Autonomous quadrotor flight despite rotor failure with onboard vision sensors: Frames vs. events. IEEE Robot. Autom. Lett. **6**(2), 580–587 (2021). https://doi.org/10.1109/LRA.2020.3048875.

162. N. Sünderhauf et al., The limits and potentials of deep learning for robotics. Int. J. Robot. Res. **37**(4–5), 405–420 (2018)

163. O. Tahri, F. Chaumette, Point-based and region-based image moments for visual servoing of planar objects. IEEE Trans. Robot. **21**(6), 1116–1127 (2005)

164. M. Tatarchenko, A. Dosovitskiy, T. Brox, Multi-view 3D models from single images with a convolutional network. Eur. Conf. Comput. Vis. (ECCV) 322–337 (2016)

165. J. Tobin et al., Domain randomization for transferring deep neural networks from simulation to the real world. IEEE/RSJ Int. Conf. Intell. Robot. Syst. (IROS) (2017)

166. E. Todorov, T. Erez, Y. Tassa, Mujoco: A physics engine for modelbased control. IEEE/RSJ Int. Conf. Intell. Robot. Syst. (IROS) 5026–5033 (2012)

167. P. Tokmakov, K. Alahari, C. Schmid, Learning motion patterns in videos. IEEE Conference on Computer Vision and Pattern Recognition (CVPR) (2017). https://doi.org/10.1109/cvpr.2017.64

168. P. Tokmakov, K. Alahari, C. Schmid, Learning video object segmentation with visual memory. IEEE International Conference on Computer Vision (ICCV) (2017). https://doi.org/10.1109/iccv.2017.480

169. J. Tordesillas, B. Thomas Lopez, J.P. How, FASTER: Fast and safe trajectory planner for flights in unknown environments, in *International Conference on Intelligent Robots and Systems (IROS, 2019)* (IEEE, 2019), pp. 1934–1940. https://doi.org/10.1109/IROS40897.2019.8968021.

170. B. Triggs et al., Bundle adjustment–a modern synthesis, in *Vision Algorithms: Theory and Practice*, vol. 1883, eds. by W. Triggs, A. Zisserman, R. Szeliski (LNCS, Springer Verlag, 2000), pp. 298–372

171. B. Ummenhofer et al., Demon: Depth and motion network for learning monocular stereo. IEEE Int. Conf. Comput. Vis. Pattern Recog. (CVPR) 5038–5047 (2017)

172. H.R. Varian, Causal inference in economics and marketing. Proceed. Nat. Acad. Sci. **113**(27), 7310–7315 (2016)

173. H. Wang, S. Xingjian, D.-Y. Yeung, Natural-parameter networks: A class of probabilistic neural networks, in *Advances in Neural Information Processing Systems* (2016), pp. 118–126

174. J.Y.A. Wang, E.H. Adelson, Representing moving images with layers. IEEE Trans. Image Proces. **3**(5), 625–638 (1994). https://doi.org/10.1109/83.334981

175. Y. Wang et al., Active co-analysis of a set of shapes. Trans. Comput. Graph. **31**(6), 165 (2012)

176. T. Wang et al., Learning nearly decomposable value functions via communication minimization, in *International Conference on Learning Representations* (2020)

177. H. Wu et al. Deep generative markov state models, in *Advances in Neural Information Processing Systems* (2018), pp. 3975–3984

178. M. Wulfmeier, I. Posner, P. Abbeel, Mutual alignment transfer learning, in *Conference on Robot Learning (CoRL)* (2017)

179. Y. Yang, G. Sundaramoorthi, S. Soatto, Self-occlusions and disocclusions in causal video object segmentation, in *Proceedings of the IEEE International Conference on Computer Vision* (2015), pp. 4408–4416

180. Y. Yang et al., Unsupervised moving object detection via contextual information separation, in *IEEE Conference on Computer Vision and Pattern Recognition, CVPR* (2019), pp. 879–888. https://doi.org/10.1109/CVPR.2019.00097.

181. T. Zhang et al., Learning deep control policies for autonomous aerial vehicles with MPC-guided policy search. IEEE Int. Conf. Robot. Autom. (ICRA) (2016)

182. Z. Zhang, D. Scaramuzza, Perception-aware receding horizon navigation for MAVs. IEEE Int. Conf. Robot. Autom. (ICRA) 2534–2541 (2018)
183. B. Zhou et al., RAPTOR: Robust and perception-aware trajectory replanning for quadrotor fast flight, in *CoRR* abs/2007.03465 (2020). arXiv: 2007.03465, https://arxiv.org/abs/2007.03465
184. B. Zhou et al., Robust and efficient quadrotor trajectory generation for fast autonomous flight. IEEE Robot. Autom. Lett. **4**(4), 3529–3536 (2019)
185. B. Zhou, P. Krähenbühl, V. Koltun, Does computer vision matter for action? Sci. Robot. **4**(30) (2019)
186. H. Zhou, B. Ummenhofer, T. Brox, DeepTAM: Deep tracking and mapping. Eur. Conf. Comput. Vis. (ECCV). 822–838 (2018)
187. T. Zhou et al., Unsupervised learning of depth and ego-motion from video. IEEE Int. Conf. Comput. Vis. Pattern Recog. (CVPR) 6612–6619 (2017)
188. Y. Zhu et al., Target-driven visual navigation in indoor scenes using deep reinforcement learning, in *International Conference on Robotics and Automation (ICRA)* (2017), pp. 3357–3364
189. Y. Zhu et al., Reinforcement and imitation learning for diverse visuomotor skills, in *Proceedings of Robotics: Science and Systems* (2018)